Contributions to Environmental Sciences & Innovative Business Technology

Contributions to Environmental Sciences & Innovative Business Technology (CESIBT) is an interdisciplinary series of peer-reviewed books dedicated to addressing emerging research trends relevant to the interplay between Environmental Sciences, Innovation, and Business Technology in their broadest sense. This series constitutes a comprehensive up-to-date interdisciplinary reference that develops integrated concepts for sustainability and discusses the emerging trends and practices that will define the future of these disciplines.

This series publishes the latest developments and research in the various areas of Environmental Sciences, Innovation, and Business Technology, combined with scientific quality and timeliness. It encompasses the theoretical, practical, and methodological aspects of all branches of these scientific disciplines embedded in the fields of Environmental Sciences, Innovation, and Business Technology.

The series also draws on the best research papers from EuroMid Academy of Business and Technology (EMABT) and other international conferences to foster the creation and development of sustainable solutions for local and international organizations worldwide. By including interdisciplinary contributions, this series introduces innovative tools that can best support and shape both the economical and sustainability agenda for the welfare of all countries, through better use of data, a more effective organization, and global, local, and individual work. The series can also present new case studies in real-world settings offering solid examples of recent innovations and business technology with special consideration for resolving environmental issues in different regions of the world.

The series can be beneficial to researchers, instructors, practitioners, consultants, and industrial experts, in addition to governments from around the world. Published in collaboration with EMABT, the Springer CESIBT series will bring together the latest research that addresses key challenges and issues in the domain of Environmental Sciences & Innovative Business Technology for sustainable development.

Bahaa Awwad · Razan Awadallah Awwad
Editors

Technology and Innovation Towards a Future of Business Sustainability

Insights in the Digital Age

Springer

Editors
Bahaa Awwad
College of Business and Finance
Ahlia University
Manama, Bahrain

Razan Awadallah Awwad
Department of Digital Business
Al-Quds University
Ramallah, Palestine, State of

ISSN 2731-8303 ISSN 2731-8311 (electronic)
Contributions to Environmental Sciences & Innovative Business Technology
ISBN 978-981-95-6468-2 ISBN 978-981-95-6469-9 (eBook)
https://doi.org/10.1007/978-981-95-6469-9

This Springer imprint is published by the registered company Springer Nature Singapore Pte Ltd.
The registered company address is: 152 Beach Road, #21-01/04 Gateway East, Singapore 189721, Singapore

Acknowledgments

The editors extend their sincere gratitude to all contributing authors for their invaluable work and insights. Special thanks are also due to Ahlia University, and Al-Quds University for their continuous support in making this volume possible.

Contents

Part I

Digital Transformation and Artificial Intelligence in Sustainability

1 The Role of Digital Transformation in Enhancing Corporate Sustainability: A Conceptual Study on Palestinian Institutions Using Digital Twin and Cloud Computing Within the ESG Framework

Eman Naser-Karajah

Abstract

This study examines the role of digital transformation in enhancing corporate sustainability within Palestinian institutions through the Environmental, Social, and Governance (ESG) framework. The study is based on a structured analytical review of relevant international and local literature, with a focus on digital transformation, ESG indicators, and the Palestinian institutional context. It identifies key structural and institutional challenges facing Palestinian organizations, including limited digital infrastructure, a shortage of digital skills, and the absence of integrated policies aligned with ESG indicators. The study demonstrates that digital transformation can support sustainability outcomes in the Palestinian context when accompanied by strategic alignment, investment in digital capacity building, and the development of national frameworks for assessing digital maturity and ESG integration. The paper offers policy-oriented and institutional recommendations to support policymakers, organizations, and development actors in advancing a resilient and sustainable digital transformation agenda.

Keywords

Digital transformation • Sustainability • ESG • Palestine • Cloud computing • Digital twin • Public policy

E. Naser-Karajah (✉)
Department of Information Technology, The Arab American University Palestine, Ramallah, Palestine
e-mail: eman.naser@aaup.edu

B. Awwad and R. A. Awwad (eds.), *Technology and Innovation Towards a Future of Business Sustainability*, Contributions to Environmental Sciences & Innovative Business Technology, https://doi.org/10.1007/978-981-95-6469-9_1

1.1 Introduction

Amid intensifying global environmental and social challenges such as climate change, resource depletion, and rising inequality, corporate sustainability has emerged as a strategic framework enabling institutions to balance economic objectives with environmental and social responsibilities (World Economic Forum 2023). This shift necessitates a transition from traditional operational models to new ones based on innovation, agility, and efficiency. At the heart of this transformation lies digital transformation: the adoption of digital technologies to enhance organizational processes, governance, and institutional performance (Aburub and Assaf 2022; Awwad and El Khoury 2024).

Among the most impactful digital tools are cloud computing, which reduces dependency on physical infrastructure, enables flexible access to data and services, lowers costs, and minimizes carbon emissions (Gokarna 2021; Mousa et al. 2024) and e-payment systems that have similarly enhanced financial performance in Palestinian banks (Awwad 2021) and the digital twin, which enables organizations to simulate real-time operations for better forecasting, improved performance, and more sustainable resource management (Ni et al. 2023).

Research increasingly confirms that the relationship between digital transformation and sustainability is becoming more evident and influential in institutional policies worldwide. Digitalization empowers organizations to track environmental impacts, enhance governance, and foster social inclusion, all in alignment with Environmental, Social, and Governance (ESG) principles (World Bank 2022; World Economic Forum 2023). These indicators are now widely used to assess institutional performance in these domains (Herzallah and Jallad 2024; Luo 2024; Awwad 2024). Integrating digital transformation into sustainability planning is a prerequisite for achieving long-term development objectives.

In the Palestinian context, however, institutions face multifaceted challenges that constrain their ability to fully capitalize on this integration. These include underdeveloped digital infrastructure, a widening digital skills gap, and the absence of supportive regulatory frameworks (ESCWA 2024; World Bank 2022; Al Najjar et al. 2023; Pizzigoni 2022). Additionally, restrictions on frequency spectrum allocation and telecommunications access limit the capacity of institutions to adopt advanced digital solutions (GSMA 2024; MAS 2023). While partial digital initiatives exist in certain sectors, a broader, strategic vision is needed, one that reflects the Palestinian context and is supported by clear benchmarks to guide consistent evaluation and development.

This study aims to examine the relationship between digital transformation and sustainability in Palestine by analyzing the contributions of technologies such as cloud computing and digital twins to corporate sustainability goals. It draws on academic literature and institutional practices to highlight both opportunities and constraints. The ultimate objective is to build a practical and context-sensitive understanding of the digital sustainability nexus, grounded in ESG standards, and to offer actionable policy recommendations for enhancing integration between technological transformation and sustainable development planning.

The importance of this research lies in its response to the persistent inability of many institutions to activate digital tools in support of sustainability objectives. Additionally, a notable gap remains in understanding the real-world impact of certain emerging technologies on ESG performance.

1.2 Methodology

This study adopts a literature-based approach to examine the relationship between digital transformation and corporate sustainability within the context of ESG indicators in Palestine. This approach is commonly employed in conceptual research to explore definitions and clarify relationships between variables by relying on a structured review of relevant literature and prior studies.

The analytical framework draws on credible international sources, including reports from organizations such as ESCWA (2024), the World Bank (2022), and the World Economic Forum (2023). It also incorporated peer-reviewed academic articles addressing topics such as digital transformation, cloud computing, digital twins, and digital infrastructure maturity in Palestine and the Arab region (Al Najjar et al. 2023; Westerman et al. 2014). Additionally, national reports from Palestinian public and private institutions working in digital governance and transformation were included (MAS 2023; GSMA 2024).

This approach is adopted in response to the limited availability of applied research in Palestine that directly addresses the connection between digital transformation and ESG indicators. Furthermore, international and local literature helps shape a comparative and adaptable theoretical understanding, especially in light of the absence of comprehensive national data on digital readiness.

To structure the literature review, thematic analysis was employed to identify key themes such as the impact of digitalization on ESG performance, the digital skills gap, governance frameworks, and the degree to which Palestinian institutions have adopted emerging technologies. These themes were then contextualized to reflect the Palestinian institutional landscape and its unique opportunities and constraints.

1.3 Literature Review

Recent studies indicate a strong and positive relationship between digital transformation and improvements in institutional performance across environmental, social, and governance dimensions. For instance, digitalization has been shown to reduce emissions, increase transparency, and strengthen stakeholder relationships (Luo 2024). Another study found that organizations that strategically adopt digital transformation see 26% higher profitability and a 12% increase in market value compared to their non-digital counterparts (Westerman et al. 2014).

In Palestine, however, reports by the World Bank show that digital infrastructure still faces deep structural challenges. Limited access to frequencies, market fragmentation, and outdated 3G and 4G networks hinder the implementation of advanced digital projects (World Bank 2022). Other local studies have highlighted that many Palestinian NGOs lack clear digital strategies, face chronic funding shortages, and suffer from limited technical human capital (Al Najjar et al. 2023).

ESCWA data reveal a significant digital skills gap in Palestine, with around 75% of adults lacking basic digital skills and fewer than 20% of institutions offering structured training for their staff (ESCWA 2024). This lack of human capital and capacity development restricts the potential for a sustainable digital transition.

Among the most promising tools for Palestinian institutions are cloud computing solutions, which offer flexibility, scalability, and cost effectiveness, particularly for resource-constrained organizations, alongside financial technology applications that support efficiency and innovation (Razia and Awwad 2023). Studies show that cloud technologies have helped SMEs in Palestine improve operational performance and reduce infrastructure costs (Mousa et al. 2024). Similarly, digital twin technology has emerged as a leading tool to enhance efficiency and reduce waste (Hamdan et al. 2021). One study found that digital twins improved building energy use and reduced operational waste through real-time monitoring and data-driven decision-making (Ni et al. 2023). This makes the technology a strong candidate for overcoming infrastructure limitations in Palestine and for helping organizations plan effectively for the future.

Nevertheless, the Palestinian context presents major obstacles to digital transformation. These include the lack of targeted research exploring the connection between digital transformation and sustainability, weak infrastructure, limited funding, and fragmented digital governance. Political and economic challenges further complicate efforts to build an integrated and sustainable digital ecosystem (ESCWA 2024; Al Najjar et al. 2023; World Bank 2022; Pizzigoni 2022).

The success of digital initiatives should not be measured solely by technological implementation. Institutions must also demonstrate the ability to integrate social and governance elements into their digital agendas. The World Economic Forum has called for the adoption of digital ESG frameworks to guide digital transformation efforts (World Economic Forum 2023), including artificial intelligence as a transformative driver of institutional performance and governance (Awwad 2024).

Despite these challenges, there is growing interest and momentum among Palestinian institutions to pursue digital innovation, especially supported by donor agencies promoting sustainability. Some institutions have already adopted advanced systems such as ERP, cloud services, and business intelligence platforms, demonstrating a clear willingness to bridge the gap between digitalization and sustainability (Mousa et al. 2024; MAS 2023).

Understanding these opportunities and barriers is a critical step toward formulating national strategies that enhance the capacity of Palestinian institutions to leverage digital transformation for sustainable development.

1.4 Exploring the Relationship Between Digital Transformation and Sustainability in the Palestinian Context

Building on the literature reviewed, this section explores how digital transformation intersects with institutional sustainability within Palestinian organizations. Findings from global case studies consistently point to a positive correlation between the adoption of digital tools such as cloud computing and digital twins and improved ESG performance (Luo 2024; World Bank 2022; Westerman et al. 2014).

Cloud computing, in particular, has emerged as a foundational pillar of digital transformation, enabling institutions to reduce operating costs and improve resource management efficiency, especially among small and medium enterprises (Gokarna 2021; Mousa et al. 2024). In parallel, digital twin technology offers organizations real-time monitoring and predictive analytics, which are especially valuable in infrastructure and energy-intensive sectors (Ni et al. 2023).

Despite these promising technologies, Palestinian institutions continue to face structural and operational obstacles that hinder digital readiness. Among the most pressing issues are the digital skills gap, the lack of structured training programs, and an absence of strategic leadership in digital transformation, particularly in the public sector (ESCWA 2024; Al Najjar et al. 2023; Aburub and Assaf 2022; Herzallah and Jallad 2024). This challenge is also evident in the banking sector, where governance plays a central role in guiding digital transformation efforts (Awwad et al. 2024).

Digital security is another critical concern. As reliance on cloud services and open platforms increases, so do the risks of cyberattacks, especially in a context where cybersecurity frameworks remain underdeveloped (Pizzigoni 2022). The lack of a national strategy that links digitalization efforts to sustainability goals has further led to fragmented and inconsistent digital initiatives, rather than a unified, long-term institutional vision (World Bank 2022; WEF 2023).

Infrastructure-related challenges are also prominent, particularly in underserved areas where digital connectivity remains weak. Poor internet penetration and outdated telecommunications systems limit equitable access to digital tools (GSMA 2024). In addition, the lack of integration across financial technology systems further restricts the digital transformation potential in key sectors (MAS 2023). Many institutions continue to rely on legacy systems that are costly and difficult to migrate to modern platforms, slowing the pace of transformation (Fahmideh et al. 2020a, b; Alkababji and Awwad 2023).

To contextualize the global findings for Palestine, Table 1.1 summarizes key studies, their focus areas, and how their insights can be applied to support Palestinian institutions in advancing digital sustainability.

Table 1.1 Summary of prior studies and their relevance to the Palestinian context

No.	References	Focus area	Key findings	Potential application in Palestine
[1]	ESCWA (2024)	Digital skills and sustainability	Severe digital gap; lack of staff training	Launch national programs to build digital capacity
[2]	World Bank (2022)	Digital transformation in Palestine	Infrastructure limitations and frequency constraints	Improve legal frameworks and digital infrastructure
[3]	Al Najjar et al. (2023)	NGOs and digital transformation	Poor planning and funding for tech projects	Provide sustainable funding for digital initiatives
[4]	Aburub and Assaf (2022)	Higher education digitization	Lack of institutional digital strategies	Standardize digital infrastructure across universities
[5]	Herzallah and Jallad (2024)	Digital leadership in the public sector	Digital leadership improves performance	Develop training programs for digital public leadership
[6]	Gokarna (2021)	Cloud adoption post-COVID	Accelerated digitization and reduced costs	Support cloud migration in the public sector
[7]	Mousa et al. (2024)	Cloud computing in Palestine	Enhanced SME efficiency	Offer technical and financial support to local SMEs
[8]	Ni et al. (2023)	Digital twins and energy management	Boosts energy efficiency in buildings	Use digital twins in municipalities and infrastructure planning
[9]	Pizzigoni (2022)	Cybersecurity in Palestine	Weak digital security systems	Embed cybersecurity in national digital strategies
[10]	Westerman et al. (2014)	Strategic digital transformation	Digitally mature firms outperform peers	Develop long-term strategic digital visions
[11]	World Economic Forum (2023)	ESG and digital governance	Need to link ESG to digitization	Create ESG-based digital transformation frameworks
[12]	Ibaid (2021)	Digital government in Palestine	Institutional slowness in adopting digital models	Accelerate digitization through a unified national roadmap
[13]	GSMA (2024)	Telecom and connectivity	Inadequate infrastructure and internet inclusion	Invest in modernizing connectivity infrastructure

(continued)

Table 1.1 (continued)

No.	References	Focus area	Key findings	Potential application in Palestine
[14]	MAS (2023)	Financial technology integration	Weak interoperability across platforms	Develop unified digital finance systems
[15]	Fahmideh et al. (2020a, b)	Legacy systems and cloud transition	Difficult data migration and high transition costs	Plan phased migration strategies to cloud systems

Source Prepared by the author (2025)

1.5 Activating Digital Transformation to Achieve Institutional Sustainability in Palestine

Based on the analysis of relevant literature and a deep understanding of the Palestinian context, it is clear that there is a pressing need for well-defined strategic actions to enable institutions to harness digital transformation as a tool for enhancing institutional sustainability, particularly within the framework of ESG standards.

The importance of these measures stems from the complex socio-political and economic realities in which Palestinian institutions operate. These challenges compel organizations to improve efficiency and resilience through the smart use of technology. To ensure effectiveness, these measures require multi-dimensional support, including legislative, financial, and institutional frameworks. Moreover, national and regional development conferences represent an opportunity to launch a "National Charter for Sustainable Digitalization" as a unifying framework to align digital transformation efforts with long-term sustainability goals. In light of this, we propose the following context-specific and actionable measures for Palestine:

1. Integrate digital transformation into national sustainable development strategies, prioritizing sectors such as education, health, energy, and public administration (Awwad and Dhair 2025).
2. Launch national digital capacity-building programs in collaboration with universities and private sector institutions to train professionals in cloud computing, open data, and digital twin technologies.
3. Provide financial and technical incentives for institutions adopting digital management systems aligned with ESG indicators.
4. Pilot the implementation of digital twin technologies in municipalities and public utilities, especially in energy, water, and transport, to reduce waste and improve operational and environmental efficiency (Ni et al. 2023).
5. Develop localized tools to assess digital quality and ESG integration, ensuring that Palestinian institutions can monitor their digital progress using relevant and practical metrics (Al Najjar et al. 2023; Aburub and Assaf 2022).

6. Enhance cybersecurity frameworks and update digital legislation to ensure national data sovereignty and secure infrastructure, especially given the restrictions on connectivity and security challenges (World Bank 2022; Pizzigoni 2022).

1.6 Monitoring the Quality of Digital Transformation

Evaluating the quality of digital transformation is essential to ensure it moves beyond being a technical process and instead becomes a meaningful driver of institutional performance and sustainability. In Palestine, where digital capacities and infrastructure are often limited, quality monitoring must rely on adaptable, low-cost, and context-sensitive tools. Globally, various models are used to assess digital progress, including:

- Digital performance indicators, which track cloud adoption, employee engagement with new systems, and the pace of transformation (World Economic Forum 2023).
- Digital maturity models, such as the MIT Digital Maturity Model or Deloitte's framework, classify institutional progress across levels from readiness to integration and innovation.

Digital ESG indicators, which connect technology use to sustainability goals—for example, reductions in carbon emissions, improved service accessibility for vulnerable groups, and greater transparency through blockchain-based systems (Razia and Awwad 2022).

However, applying these international models in Palestine presents challenges. Many assume a robust digital environment and adequate financial and human resources, which may not always be available locally. Therefore, developing simple and locally tailored assessment tools is a more realistic and effective approach. These may include:

- Internal surveys to assess staff satisfaction with digital systems;
- Monitoring reductions in paper or electricity usage after system digitization;
- Basic metrics tracking digital service response times or system uptime (ESCWA 2024; World Bank 2022).

These tools offer a practical understanding of the real impact of digital transformation and provide policymakers with evidence-based insights. Most importantly, they allow Palestinian institutions to gradually and realistically track digital progress, adapting strategies based on measurable outcomes rather than assumptions. In this way, digital transformation can align with sustainability objectives even in a resource-constrained environment.

1.7 Political and Strategic Guidelines

Based on an in-depth analysis of the Palestinian context, its digital transformation challenges, and potential opportunities, this study underscores the necessity of adopting an integrated strategic framework. Such a framework should promote the role of digitalization in achieving institutional sustainability while requiring a holistic intervention that addresses legislative reform, infrastructure enhancement, and human capacity development. All these efforts must be tailored to the complex political and economic conditions specific to Palestine.

Accordingly, the study offers a set of key recommendations to support the implementation of digital transformation in Palestinian institutions in alignment with ESG goals:

1. Develop a comprehensive national strategy for sustainable digital transformation, with a strong focus on integrating ESG indicators across public policies and strategic plans.
2. Strengthen digital infrastructure in various regions, particularly through securing external funding and forming international partnerships, given the limitations in domestic investment due to ongoing financial crises.
3. Launch national capacity-building programs to enhance the digital competencies of public and private sector employees, as well as university students, especially in cloud computing, open data, and digital twin applications.
4. Provide financial and technical support to small and medium-sized enterprises (SMEs) to adopt cloud-based platforms and smart management tools that can enhance operational performance and efficiency.
5. Design national frameworks for evaluating digital quality, closely linked to sustainability metrics and supported by tools that enable periodic progress monitoring.
6. Establish dedicated national units within government structures tasked with tracking the quality of digital transformation, using ESG-based indicators to ensure meaningful and measurable progress.

1.8 Conclusion

This study explored how digital transformation can serve as a key enabler of institutional sustainability in Palestine. Drawing on both international and local literature, the research confirmed that technologies like cloud computing and digital twins do more than improve operational efficiency; they also play a critical role in advancing environmental, social, and governance performance. These tools offer institutions greater agility and adaptability, particularly in environments marked by economic and political volatility.

The study highlighted major challenges specific to the Palestinian context, such as weak infrastructure, a significant digital skills gap, and the absence of comprehensive institutional strategies. These barriers limit the effective deployment of digital technologies. It also stressed the urgent need to strengthen cybersecurity and develop legal frameworks that ensure data protection, essential steps toward building a secure and resilient digital environment.

To bridge the gap between potential and practice, the study presented a series of context-sensitive, actionable recommendations. Chief among them was the call for a clear national strategy that positions digital transformation as a core component of sustainable development. The recommendations emphasized the importance of investing in digital skills, supporting SMEs, and creating reliable tools for measuring the real impact of digitalization on institutional performance.

Ultimately, the value of this research lies in its policy-oriented and conceptual contribution to understanding digital transformation and sustainability in the Palestinian context. It sheds light on the current realities of Palestinian institutions, offering realistic and locally grounded solutions. These insights can guide policymakers in taking deliberate and informed steps toward building a more sustainable and digitally empowered future, one that enhances institutional capacity to meet present challenges and seize future opportunities.

References

Aburub I, Assaf D (2022) Digital transformation in higher education in Palestine. Balt J Law Polit
Al Najjar MT, Al Shobaki MJ, El Talla SA (2023) Digital transformation in Palestinian NGOs. Int J Eng Inf Syst (IJEAIS)
Alkababji MW, Awwad BS (2023) Big data tools and techniques in developing accounting practices and the impact of their analysis on the financial performance. Cent Eur Manag J 31(1):618–631
Awwad BS (2021) The role of e-payments in enhancing financial performance: a case study of the Bank of Palestine. Banks Bank Syst 16(4):114
Awwad B (ed) (2024) The AI revolution: driving business innovation and research. Springer
Awwad BSAL (2024) Governance with relationship between entrepreneurship and economic growth in Palestine. Int J Law Manag 66(2):259–287
Awwad BS, Dhair MA (2025) The impact of implementing digital transformation on the quality of government service provision in the Palestinian Ministry of Higher Education and Scientific Research. In: Sustainability in light of governance and artificial intelligence applications, p 71
Awwad B, El Khoury R (2024) Information technology governance and bank performance: evidence from Palestine. J Decis Syst 33(2):311–334
Awwad BS, Razia BS, Razia AS (2024) Digital transformation under the governance of Palestinian banks. Discov Sustain 5(1):76
ESCWA (2024) Annual SDG review 2024: skills & innovation in Arab region
Fahmideh M, Daneshgar F, Beydoun G, Rabhi F (2020a) Challenges in migrating legacy software systems to the cloud: an empirical study. arXiv
Fahmideh S, Daneshgar F, Beydoun G (2020b) Legacy system challenges in digital transformation: a socio-technical perspective. Inf Syst Front 22(2):395–414. https://doi.org/10.1007/s10796-018-9873-8
Gokarna M (2021) Cloud adoption post-COVID. arXiv
GSMA (2024) Digital 2024: Palestine
Hamdan A, Hassanien AE, Khamis R, Alareeni B, Razzaque A, Awwad B (eds) (2021) Applications of artificial intelligence in business, education and healthcare, vol 954. Springer, Cham
Herzallah AM, Jallad A (2024) Role of digital leadership in Palestinian government. J Soc Sci (JSS) 30(4)

Luo X (2024) Digital transformation and ESG performance: evidence from China
MAS (2023) Digital financial services roadmap
Mousa K, Zhang Z, Sumarliah E, Hamdan IKA (2024) Cloud computing adoption among SMEs in Palestine. Int J Innov Inf Technol (IJIIT)
Ni Z, Zhang C, Karlsson M, Gong S (2023) Leveraging digital twins for energy efficiency. arXiv
Pizzigoni S (2022) Cybersecurity & digitalization in Palestine. Labiba
Razia B, Awwad B (2022) A comprehensive review of blockchain technology and its related aspects in higher education. In: Technologies, artificial intelligence and the future of learning post-COVID-19: the crucial role of international accreditation, pp 553–571
Razia B, Awwad B (2023) Review of financial technology applications and their related aspects. In: From industry 4.0 to industry 5.0. Springer, Cham, pp 893–902
Westerman G, Bonnet D, McAfee A (2014) Leading digital: turning technology into business transformation. Harvard Business Review Press
World Bank (2022) Digital transformation for Palestinian economic growth
World Economic Forum (2023) Measuring digital sustainability

Digital Accounting and Innovation as Drivers of Business Sustainability

2

Bahaa Awwad, Raed Abdelhaq, and Ammar Zakaria Salem

Abstract

This study addresses the strategic role of digital accounting in promoting business sustainability amidst emerging technologies such as artificial intelligence, blockchain, the Internet of Things, and big data analytics. Following international environmental, social, and corporate governance (ESG) standards such as the GRI and SASB, digital accounting systems enable real-time monitoring, automated reporting, and improved decision-making. The study addresses first-line implementation plans, performance measures, industry deviations, and potential risks due to over-digitization. Case studies and theoretical descriptions provide scientific and applied knowledge about the future direction of guaranteeing ESG, digitization, and bundled reporting. The study concludes with practical recommendations and research directions for promoting ethical and scalable application of digital bookkeeping to achieve sustainability in various organizational and regulatory contexts.

B. Awwad (✉)
College of Business and Finance, Ahlia University, Manama, Bahrain
e-mail: bawwad@ahlia.edu.bh

R. Abdelhaq
Accounting Department, Faculty of Business and Communications, An-Najah National University, Nablus, Palestine
e-mail: raed.abdelhaq@najah.edu

A. Z. Salem
Department of Administrative and Financial Sciences, Palestine Technical University—Kadoorie, Tulkarm, Palestine
e-mail: ammar.salem@ptuk.edu.ps

B. Awwad and R. A. Awwad (eds.), *Technology and Innovation Towards a Future of Business Sustainability*, Contributions to Environmental Sciences & Innovative Business Technology, https://doi.org/10.1007/978-981-95-6469-9_2

Keywords

Digital accounting • ESG reporting • Artificial intelligence • Blockchain • Sustainability assurance • Real-time monitoring • GRI • SASB • Integrated reporting • Digital transformation • Ethical governance

2.1 Introduction

Sustainability has become an essential aspect of the global business landscape, as companies are held more accountable by regulators, stakeholders, and investors to provide evidence of their environmental, social, and governance (ESG) conduct (Ahmad et al. 2024; Awwad et al. 2023; Awwad and Hamdan 2018). To this end, organizations are adopting increasingly computerized accounting systems with advanced emerging technologies, including artificial intelligence (AI), blockchain, and real-time analytics, to enhance the efficacy and credibility of sustainability reporting. Computerized accounting is also taking increasing significance in monitoring environmental, social, and governance (ESG) performance via digital data collection, real-time reporting, and enhanced decision-making. For instance, artificial intelligence solutions enhance the accuracy of carbon footprint disclosure and ESG forecasting processes (De Silva et al. 2025; Razia et al. 2024; Awwad et al. 2024a). Blockchain supports the integrity of sustainability reporting with tamper-proof records and transparent supply chain monitoring (Nwariaku et al. 2024; Razia and Awwad 2022). Similarly, IoT sensors and cloud computing allow firms to monitor environmental indicators in real time, closing the gap between sustainability and business operations. The existing empirical evidence demonstrates that digital transformation has a significantly positive effect on ESG performance, especially where there is alignment between digital technologies and corporate sustainability strategies (Kwilinski et al. 2023; Hamdan et al. 2021; Awwad 2024a). When appropriately utilized, digital accounting transforms from a simple compliance tool into an enabler of sustainable business resilience and sustainable value creation.

This study explores the intersection of digital accounting and innovation as enablers of sustainable business practices. It highlights key technologies, strategic approaches to implementation, and ways to measure their impact on ESG performance.

2.2 The Role of Digital Accounting in Sustainable Business

2.2.1 From Traditional to Digital Accounting

Cloud accounting is a move away from paper-based manual processes to cloud-based automated ones with access to financial and non-financial data. Differing from the classical delay-based accounting approach, digital solutions allow organizations to monitor performance in real time and respond. Automation provides more accuracy at lower human touch, and cloud-based solutions provide increased accessibility across departments and geographies. These skills are especially important when considering ESG, in which timely data gathering is very important in maintaining the pace of reporting requirements and risk management. Digital accounting thus allows businesses to shift their financial activities toward more comprehensive sustainability agendas (Abdelhaq et al. 2025; Awwad et al. 2024b).

2.2.2 Integrating Sustainability into Accounting Systems

Modern accounting systems increasingly incorporate sustainability measurements into their fundamental functionality. The majority of ERP systems nowadays include independent carbon emission tracking, social impact measurement, and sustainability reporting modules based on such standards as the Global Reporting Initiative (GRI). When these tools are integrated, companies can generate ESG reports automatically, reducing the reliance on independent, manual procedures. Besides efficiency improvement, it ensures conformity between financial and non-financial reports. Also, dashboards for sustainability deliver real-time monitoring of ESG performance to enable corporations to review their performance and adjust strategies accordingly. This incorporation boosts accountability and supports efficient decision-making across all organizational levels (Abdeljawad et al. 2025; Salah and Awwad 2024; Awwad 2024b).

2.3 Key Innovations Supporting Digital Accounting and Sustainability

2.3.1 Artificial Intelligence (AI) in Financial Analysis

AI-based tools play a pivotal role in today's financial and ESG analysis. AI-based tools enhance sustainability forecasting, fraud detection, and analysis of trends relating to ESG. Machine learning algorithms can process environmental risk factors in investment portfolios or assign financial and natural capital to reduce waste. Current studies reaffirm that AI enables improved ESG performance by predictive analysis and data-driven automations (Li et al. 2025). AI, in finance, not only contributes to efficiency but also helps firms become proactive and data-driven in sustainability strategies.

2.3.2 Blockchain for ESG Assurance

Blockchain technology is increasingly used to ensure ESG disclosures are more transparent and reliable. It possesses unalterable records of economic and sustainability-related transactions, reducing the scope for greenwashing or tampering. Use of smart contracts also makes automated compliance with ESG easier, such as issuance of payments for verified sustainability milestones (Nguyen et al. 2025). According to a recent empirical study, blockchain increases investor confidence in ESG data with traceability and verifiability (Gu et al. 2022; Abdelhalim and Hassan 2025). It is therefore a significant assurance and accountability tool in sustainability reporting.

2.3.3 Internet of Things (IoT) for Real-Time Accounting Data

The integration of IoT sensors into accounting systems has enabled the tracking of environmental information in real time, such as energy consumption, emissions, and water consumption. The sensors provide ongoing information to accounting systems, which enable dynamic reporting on ESG and quick response to inefficiencies or deviations (Petcu et al. 2024; WSJ/Deloitte 2024). For instance, overconsumption of energy is identified automatically, enabling instant correction. All such real-time information increases internal control as well as external transparency and, therefore, makes IoT a significant enabler of operational efficiency and cost-effectiveness.

2.3.4 Big Data and Sustainability Dashboards

Big data analytics enables thorough ESG monitoring by bringing together big volumes of structured and unstructured data across departments. Firms increasingly rely on sustainability dashboards to track key indicators in real time and compare performance over time and between departments. Recent studies in the Saudi industrial economy confirmed that the combination of big data with risk management led to measurable improvements in sustainability performance (Abdelhalim and Hassan 2025; Alkababji and Awwad 2023). These dashboards support financial professionals and sustainability officers to make proactive, data-driven decisions and foster a culture of continuous ESG improvement, as shown in Fig. 2.1.

Digital Technologies and Their Contributions to ESG Dimensions

	Environmental	Social	Governance
Artificial Intelligence	AI: Emission forecasting	AI: Labor analytics	AI: Fraud detection
Blockchain	Blockchain: Carbon tracking	Blockchain: Ethical sourcing	Blockchain: Audit trails
IoT	IoT: Energy monitoring	IoT: Workplace safety monitoring	IoT: Compliance alerts

Fig. 2.1 Digital technologies and their contributions to ESG dimensions. *Source* Prepared by the authors

2.4 Implementation Strategies for Sustainable Digital Accounting

2.4.1 Aligning with ESG and Regulatory Frameworks

To ensure consistency and transparency in sustainability reporting, digital accounting systems must align with leading global standards such as GRI, SASB, and the EU's CSRD. GRI's 2025 digital taxonomy allows ESG data to be machine-readable and standardized across platforms (Global Reporting Initiative [GRI] 2025). Likewise, SASB standards support industry-specific ESG disclosures that integrate smoothly into financial systems (SASB 2024). These frameworks help firms meet regulatory expectations and improve the comparability of non-financial data.

2.4.2 Building Interdisciplinary Teams

The integration of ESG into digital accounting requires collaboration between finance, technology, and sustainability teams. Bridging the skills gap is essential, as many professionals lack the expertise to handle ESG data within digital systems (FM Magazine 2022). Companies such as Bain & Company advocate for structured upskilling to embed ESG in core operations and strategy (Bain & Company 2024). Interdisciplinary collaboration ensures accurate ESG reporting and meaningful data interpretation.

2.4.3 Piloting Digital Solutions

A pilot-based approach allows firms to test digital ESG tools on a small scale before scaling up. This reduces risk and builds internal support. Deloitte (2024) emphasizes starting with focused solutions—such as ESG dashboards or automated tagging—then expanding based on outcomes. Pilots enhance flexibility, improve adoption rates, and help organizations refine sustainability data processes efficiently.

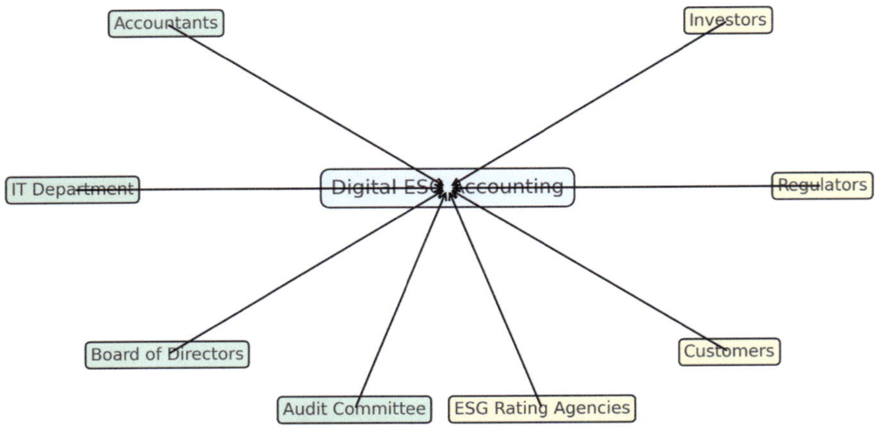

Fig. 2.2 Stakeholder ecosystem in digital ESG accounting. *Source* Prepared by the authors

2.4.4 Leveraging Cloud-Based Accounting Platforms

Cloud platforms such as SAP are integrating ESG modules into core accounting workflows to support audit-ready sustainability reporting. SAP Fioneer's ESG KPI Engine offers real-time tracking of key environmental metrics within ERP systems (SAP Fioneer 2024). These tools minimize manual work and embed ESG into financial operations, enabling firms to meet growing assurance demands and improve data accuracy.

2.4.5 Mapping Key Stakeholders in Digital ESG Integration

Effective digital ESG implementation requires coordination among a wide range of internal and external stakeholders. Figure 2.2 illustrates the ecosystem of actors involved in shaping, using, and evaluating digital sustainability accounting systems.

2.5 Measuring the Impact of Digital Accounting on Sustainability

2.5.1 Key Metrics

Digital accounting supports the use of standardized ESG metrics such as carbon emissions per revenue unit, energy efficiency ratios, and social investment percentages. These indicators enable consistent tracking and benchmarking of sustainability performance. By integrating them into accounting systems, firms enhance transparency and decision-making (Abdelhalim and Hassan 2025).

2.5.2 Real-Time Monitoring and Auditing

IoT and AI technologies allow for real-time data collection and ESG audit trails. This ensures continuous verification of environmental performance and improves internal controls. Organizations benefit from faster responses to anomalies and more reliable disclosures (Nguyen et al. 2025).

2.5.3 Integration into Financial Reporting

Many firms now embed sustainability indicators directly into financial reports to meet investor and regulatory expectations. This reflects a shift toward integrated reporting where ESG data is treated as financially material. Embedding these metrics enhances report quality and stakeholder trust (De Silva et al. 2025).

2.6 Case Studies

Deloitte has implemented blockchain in audit processes to ensure more accurate ESG disclosures and reduce manipulation risks (WSJ/Deloitte 2024). Microsoft uses AI tools to calculate internal carbon pricing, encouraging lower emissions across departments (Li et al. 2025). PwC provides clients with ESG digital platforms that automate compliance and improve sustainability reporting (Petcu et al. 2024).

2.7 Challenges in Digital Accounting for Sustainability

- **Data Reliability**: Inconsistent ESG data across departments weakens report accuracy and credibility. Integrating sustainability metrics into digital systems can solve this by standardizing data sources (Petcu et al. 2024).
- **Cost of Implementation**: High initial investment in ESG digital systems can be a barrier, especially for SMEs. Despite this, long-term benefits in efficiency and compliance often justify the cost (SAP Fioneer 2024).
- **Skills Gap**: A shortage of professionals skilled in both digital accounting and ESG analytics limits implementation. Bridging this gap requires targeted upskilling programs and digital training (FM Magazine 2022).
- **Regulatory Uncertainty**: Constantly evolving ESG standards create ambiguity in reporting practices. Adopting frameworks like GRI or SASB helps ensure consistency and reduce compliance risks (GRI 2025).

2.8 Digital Assurance and the Future of ESG Audit

The way forward for ESG assurance lies in the use of digital technologies such as blockchain and AI to foster transparency and verifiability. Blockchain provides tamper-evident records that can be used to audit sustainability claims, thereby reducing the risk of greenwashing and manipulation (Gu et al. 2022). Meanwhile, AI tools provide automated audit trails and anomaly detection within ESG data, improving audit efficiency and integrity (Nguyen et al. 2025). As ESG data matures, digital assurance frameworks will be essential to maintain stakeholder trust and regulatory compliance (De Silva et al. 2025).

2.9 Influence of National and Sectoral Contexts on Digital ESG Practices

The adoption of digital ESG practices varies significantly across national regulatory environments and industry segments. National environments characterized by developed digital infrastructure and ESG regulations, e.g., in the EU, are leading the way in embracing advanced reporting tools (GRI 2025). Sectoral factors have an influence as well; e.g., energy and manufacturing firms are more likely to use IoT and real-time reporting due to high environmental exposure (Petcu et al. 2024). It is necessary to understand such contextual forces in order to fit digital accounting projects to specific organizational settings (Abdelhaq et al. 2025).

2.10 Opportunities and Risks of Over-Digitalization in Sustainability Reporting

While digitalization enhances ESG reporting, its over-reliance on automated systems can introduce risks of loss of human judgment, algorithmic bias, or overly complex disclosures. Although AI tools can help increase speed and scale of analysis, without appropriate governance, they may compromise transparency or accuracy (Li et al. 2025). Stakeholders can also find it difficult to interpret highly technical dashboards or digital-only reports. Balancing technological advancement with ethical and interpretive safeguards remains critical for ensuring the reliability and inclusiveness of sustainability information (Kwilinski et al. 2023).

2.11 Summary Table: Digital Technologies and ESG Dimensions

Table 2.1 summarizes how selected digital technologies support the environmental, social, and governance (ESG) pillars of sustainability.

Table 2.1 Digital technologies and ESG dimensions

Technology	Environmental impact	Social impact	Governance enhancement
Artificial intelligence	Forecasts carbon emissions; optimizes energy use	Monitors labor trends and safety risks	Detects fraud; enables automated controls
Blockchain	Tracks carbon credits and energy use	Verifies ethical sourcing in supply chains	Provides immutable audit trails
Internet of Things (IoT)	Monitors emissions and energy in real time	Tracks health and safety conditions	Enables real-time compliance oversight

Source Li et al. (2025), Gu et al. (2022) and Nguyen et al. (2025)

2.12 Comparative Appendix: Global ESG Reporting Standards

A cursory examination of major ESG disclosure frameworks can inform readers about the regulations that influence digital accounting systems.

- GRI (Global Reporting Initiative): Utilized by numerous; committed to being accessible to all. Assists all sectors in publishing full ESG data and increasingly digital (GRI 2025).
- SASB (Sustainability Accounting Standards Board): Investor-focused; prioritizes materiality by industry (SASB 2024).
- CSRD (Corporate Sustainability Reporting Directive—EU): Legally binding; double materiality and machine-readable sustainability reporting mandated from 2023 (GRI 2025).

2.13 Sectoral and National Contexts: A Regional Lens

Digital ESG practices are shaped by national regulations as well as sector-specific industry needs. As an illustration, firms in highly regulated sectors (such as manufacturing or energy) would typically set up real-time ESG data systems using IoT or blockchain technologies (Petcu et al. 2024). In the MENA region, poor infrastructure and patchy regulations would hamper adoption, but in Palestine, there is evidence of nascent convergence with international frameworks (Abdelhaq et al. 2025).

2.14 Critical Reflections: Balancing Innovation and Practicality

While digital technology carries revolutionary potential, over-reliance on AI and automation in ESG reporting is a concern. For example, black boxes of algorithms or poorly trained AI models may produce biased or false sustainability scores (Li et al. 2025). Further, implementation costs that are too high may preclude SMEs and developing economy organizations (Kwilinski et al. 2023). Hence, technological integration needs to be aligned with ethical regulation and local capacity issues.

2.15 Future Research Questions in Digital Accounting and ESG

To encourage academic advancement, future researchers should explore the following questions: Ahmad et al. (2024), De Silva et al. (2025) and Nwariaku et al. (2024).

- How does organizational culture moderate the impact of digital accounting tools on ESG reporting quality?
- What risks emerge from the use of generative AI in corporate sustainability reporting?
- How can blockchain be adapted for ESG auditing in low-resource or emerging market contexts?

2.16 Recommendations

- **Adopt Integrated Digital ESG Frameworks**: Firms should embed ESG metrics into digital accounting platforms aligned with globally recognized standards such as GRI and SASB. This enhances consistency, comparability, and decision-usefulness in sustainability reporting (GRI 2025; SASB 2024).
- **Invest in Workforce Upskilling**: Organizations must prioritize digital and ESG-specific training programs for accountants and sustainability professionals. Upskilling is essential to address talent gaps and ensure accurate ESG integration (FM Magazine 2022; Bain & Company 2024).
- **Pilot Digital Solutions Before Scaling**: A phased, low-risk approach using pilot projects—such as ESG dashboards or blockchain assurance tools—allows firms to assess digital ESG systems before organization-wide adoption (Deloitte 2024).
- **Balance Innovation with Ethical Oversight**: To avoid over-digitalization risks, firms should implement governance frameworks that ensure transparency, explainability, and accountability in AI and blockchain systems (Li et al. 2025; Kwilinski et al. 2023).

- **Tailor Digital Accounting to Sectoral and Regional Contexts**: Digital ESG strategies should consider sector-specific needs and regulatory environments, particularly in high-impact industries or developing markets (Petcu et al. 2024; Abdelhaq et al. 2025).
- **Leverage Real-Time Monitoring Technologies**: The adoption of IoT-enabled platforms enhances sustainability performance monitoring and audit preparedness through continuous data collection and automated analysis (Nguyen et al. 2025).
- **Stay Aligned with Regulatory Developments**: Firms must remain adaptive to evolving ESG regulations by aligning their digital tools with emerging frameworks such as CSRD and GRI's digital taxonomy (GRI 2025).
- **Encourage Cross-Sector Research and Collaboration**: Researchers should continue to explore digital ESG implementation across organizational contexts. Suggested research questions include: How does organizational culture influence digital ESG performance? What risks does generative AI pose in ESG reporting? How can blockchain be adapted for ESG assurance in developing markets? (Ahmad et al. 2024; De Silva et al. 2025; Nwariaku et al. 2024).

2.17 Discussion and Conclusion

This study examined the transformative role of digital accounting technologies—such as artificial intelligence, blockchain, IoT, and big data—in driving sustainability-oriented business practices. The findings demonstrate that digital accounting enhances transparency, improves real-time ESG monitoring, and facilitates regulatory compliance. However, its impact is significantly shaped by sectoral dynamics, organizational capabilities, and national contexts, underscoring the importance of tailored implementation strategies (Petcu et al. 2024; Abdelhaq et al. 2025).

Moreover, while digitalization offers unprecedented opportunities in ESG reporting, it also introduces risks, including loss of human oversight, algorithmic bias, and over-dependence on automated systems (Li et al. 2025). Therefore, firms must implement ethical and governance safeguards to ensure that these technologies enhance rather than undermine accountability and trust (Kwilinski et al. 2023).

Future academic efforts should focus on bridging theoretical and practical gaps, especially in under-researched contexts such as SMEs and emerging economies. The continued integration of ESG into financial systems requires not only technological innovation, but also strong regulatory frameworks, inclusive digital skills, and a balanced approach to data governance.

In conclusion, digital accounting is not merely a tool for operational efficiency—it is a strategic instrument for enabling sustainability, fostering stakeholder trust, and ensuring long-term business resilience in an increasingly ESG-driven world.

References

Abdelhalim AM, Hassan M (2025) Exploring the impact of big data analytics and risk management convergence on sustainability performance development: an accounting perspective. Discov Sustain 6(1):175

Abdelhaq R, Salem A, Rabaia DJ, Jardaneh L, Nour AI, Al-Sartawi AMM (2025) Corporate governance and intellectual capital efficiency: empirical evidence from Palestine. Int J Intellect Prop Manag 15(2):180–203

Abdeljawad I, Rabaia DJ, Abdelhaq R, Salem A (2025) The association between corporate social responsibility disclosure and dividend policy: insights from an emerging economy. Int J Disclosure Gov 1–19

Ahmad H, Yaqub M, Lee SH (2024) Environmental-, social-, and governance-related factors for business investment and sustainability: a scientometric review of global trends. Environ Dev Sustain 26(2):2965–2987

Alkababji MW, Awwad BS (2023) Big data tools and techniques in developing accounting practices and the impact of their analysis on the financial performance. Cent Eur Manag J 31(1):618–631

Awwad B (ed) (2024a) The AI revolution: driving business innovation and research: volume 1, vol 524. Springer Nature

Awwad BSAL (2024b) Governance with relationship between entrepreneurship and economic growth in Palestine. Int J Law Manag 66(2):259–287

Awwad BSA, Hamdan AMM (2018) Gulf cooperation council energy sectors governance and dividend policy. Int J Econ Financ Issues 8(4):163

Awwad BS, Alkababji MW, Zidan S (2023) Governance and social responsibility: Palestinian investment fund. Bus Econ Res 13(4):185–215

Awwad BS, Alkababji MW, Razia BS (2024a) Artificial intelligence and the quality of accounting information in Palestinian industrial companies. J Financ Rep Account (ahead-of-print)

Awwad BS, Razia BS, Razia AS (2024b) Digital transformation under the governance of Palestinian banks. Discov Sustain 5(1):76

Bain & Company (2024) Embedding ESG in strategy and operations: upskilling for transformation. https://www.bain.com

De Silva P, Gunarathne N, Kumar S (2025) Exploring the impact of digital knowledge, integration and performance on sustainable accounting, reporting and assurance. Meditari Account Res 33(2):497–552

Deloitte (2024) Starting small, scaling fast: how to pilot ESG digital solutions effectively. https://www2.deloitte.com

FM Magazine (2022) Bridging the ESG skills gap in finance and accounting. https://www.fm-magazine.com

Global Reporting Initiative (GRI) (2025) GRI launches new digital taxonomy to improve ESG disclosures. https://www.globalreporting.org

Gu Y, Jiang L, Yu W, Dai J (2022) Towards blockchain-enabled ESG reporting and assurance: from the perspective of P2P energy trading. SSRN Electron J

Hamdan A, Hassanien AE, Khamis R, Alareeni B, Razzaque A, Awwad B (eds) (2021) Applications of artificial intelligence in business, education and healthcare, vol 954. Springer, Cham

Kwilinski A, Lyulyov O, Pimonenko T (2023) Unlocking sustainable value through digital transformation: an examination of ESG performance. Information 14(8):444

Li J, Wu T, Hu B, Pan D, Zhou Y (2025) Artificial intelligence and corporate ESG performance. Int Rev Financ Anal 102:104036

Nguyen NM, Abu Afifa MM, Thi Truc Dao V, Van Bui D, Vo Van H (2025) Leveraging artificial intelligence and blockchain in accounting to boost ESG performance: the role of risk management and environmental uncertainty. Int J Org Anal

Nwariaku H, Fadojutimi B, Lawson LGL, Agbelusi J, Adigun OA, Udom JA, Olajide TD (2024) Blockchain technology as an enabler of transparency and efficiency in sustainable supply chains. Int J Sci Res Arch 12(2):1779–1789

Petcu MA, Sobolevschi-David MI, Curea SC (2024) Integrating digital technologies in sustainability accounting and reporting: perceptions of professional cloud computing users. Electronics 13(14):2684

Razia B, Awwad B (2022) A comprehensive review of blockchain technology and its related aspects in higher education. In: Technologies, artificial intelligence and the future of learning post-COVID-19: the crucial role of international accreditation, pp 553–571.

Razia AS, Alkababji MW, Ruzieh AS, Awwad BS, Razia BS (2024) Competitive advantage in light of AI to achieve innovative as a key of the SDGS: Palestinian insurance industry. J Lifestyle SDGs Rev 4(1):e01870

Salah AS, Awwad BS (2024) A theoretical review of artificial intelligence and tax compliance. In: Achieving sustainable business through AI, technology education and computer science: volume 3: business sustainability and artificial intelligence applications, pp 53–64

SAP Fioneer (2024) ESG KPI engine for SAP ERP: a new approach to embedded sustainability. https://www.sapfioneer.com

SASB (2024) SASB standards overview: industry-based ESG disclosure. https://www.sasb.org

WSJ/Deloitte (2024) How connectivity technologies can unlock sustainability value

3

Leveraging Artificial Intelligence in Digital Marketing to Drive Sustainable Consumer Behavior: Pathways and Perceptions for Supply Chain Sustainability

Dima A. Sawaftah

Abstract

The growing importance of ecological and social crises has moved sustainability to an essential situation in international debates, affecting business approaches and customer choices. In this rapidly evolving setting, AI-powered digital marketing provides a unique possibility to boost sustainable consumer behavior and enhance the sustainability of the supply chain. While existing literature has studied the different uses of AI-powered marketing, sustainable customer behavior, and the sustainability of supply chain management, an integrated conceptual structure that merges these interconnected arenas has not yet substantially studied. This chapter proposes advanced theories to address important research gaps, offering new paths for understanding and promoting sustainable consumption through AI-powered digital marketing, with an explicit emphasis on transparency and ethical actions over the supply chain. The consequences of this chapter will benefit both academic and practical attention.

Keywords

AI • Digital marketing • Sustainability • Consumer behavior • Supply chain

D. A. Sawaftah (✉)
Arab American University, Ramallah, Palestine
e-mail: dima.sawaftah@aaup.edu

B. Awwad and R. A. Awwad (eds.), *Technology and Innovation Towards a Future of Business Sustainability*, Contributions to Environmental Sciences & Innovative Business Technology, https://doi.org/10.1007/978-981-95-6469-9_3

3.1 Introduction

3.1.1 The Authority of Sustainable Customer Behavior in the Digital Age

The growing ecological emergency and rising social consciousness have highlighted the significant requirement for sustainable activities in all fields (Zabaniotou 2020). In the commerce era, an important move is proceeding, shifting away from out-of-date profitable options to focus on ecological preservation and ethical behavior. This model modification is mainly obvious in customer behavior, as a greater number of people are looking for goods and services that are associated with their sustainable principles (Purcărea et al. 2022). Sustainable customer behavior, including activities such as buying ecological goods, decreasing usage, and participating in reprocessing, has moved from a minority issue to a common practice. The age of digitization, described by extensive interaction and huge changes in information, intensifies both the obstacles and possibilities in enhancing these types of actions (Rachinger et al. 2019). Online media may promote subjective utilization forms throughout owing to their availability and customized marketing; they additionally present unparalleled chances for encouraging eco-friendly adoptions and growing awareness regarding ecological implications. Purchasing eco-friendly products, consumption reduction, and recycling engagement are some of the actions that shape sustainable customer behavior and are now as mainstream expectations (Yadav and Pathak 2017). Through personalized marketing and enhanced access to Digital platforms, it can simplify and provide unique avenues for marketing both unsustainable and sustainable alternatives and expand environmental influence awareness (Theodorakopoulos and Theodoropoulou 2024).

3.1.2 The Artificial Intelligence Role Within Digital Marketing

The use of Artificial Intelligence (AI) has grown into a revolutionary force throughout many sectors, particularly digital marketing (Rasheed et al. 2020). Technologies guided by AI and systems are developing the way in which organizations communicate with customers, investigate data, and improve promotion approaches (Mossavar-Rahmani and Zohuri 2024). AI offers outstanding potential for improving customer engagement and boosting revenue using analytical models which predict customers' needs and customize recommendations that personalize choices of products (Ruswandi and Khaeru Ahmadi 2023). In the era of digital marketing, the capability of AIs to examine and explain huge data sets allows businesses to get extensive knowledge about customer desires, enhance marketing effectiveness, and simplify repetitive responsibilities (Ziakis and Vlachopoulou 2023). The use of this invention acts as a two-sided instrument for a sustainable future: whereas it may be utilized to encourage sustainable goods and actions, it also transmits the risk of intensifying unmanageable use if not handled legally and reliably. This revolutionized technology, if not implemented in an ethical and responsible way, creates a multifaceted sustainability challenge: it can be employed to market environmentally friendly products; on the other hand, it can also exacerbate existing risks that intensify unsustainable consumption (Pappas et al. 2023).

3.1.3 Bridging the Gaps Between AI-Driven Digital Marketing and Supply Chain Sustainability

The elaborated nexus between AI-driven digital marketing, sustainable consumer behavior, and the sustainable supply chain is a complicated yet critical topic of investigation. A supply chain sustainability chain manages environmental, social, and economic influences and guarantees good domination applies through the growth of goods and services. AI-driven digital marketing can execute as an influential channel in this ecosystem, customer demand prompting for sustainable products and, in sequence, steering businesses to implement more sustainable supply chain attempts. The transparency of the supply chain might be increased through AI by monitoring products as they move from the point of origin to the landing place, serving customers with full information about proper sourcing and ecological impact. It is aided by channels within digital marketing, which can let customers make decisions, thus putting pressure on businesses to advance their supply chain sustainability. In spite of the salient underlying, there is a serious gap in the definite implementations and paths by which AI-driven digital marketing can effectually nurture sustainable consumer behavior that truthfully affects the sustainability of the supply chain.

3.1.4 Chapter's Objectives

This chapter aims to carefully investigate the nexus between AI, digital marketing, sustainable consumer behavior, and supply chain sustainability. Precisely, it strives for:

(i) Classify and investigate the current research gaps across the disciplinary approach.
(ii) Recommend and utilize innovative theoretical lenses to achieve a greater understanding of the multifaceted relations.
(iii) Identify the paths over which AI-powered digital marketing can enhance sustainable consumer choices and contribute to supply chain sustainability.
(iv) Debate the ESM consequences of using AI for sustainability in digital marketing.

The rest of this chapter is as follows: Sect. 3.2 examines the theoretical foundations and underlines the realized research gaps, offering advanced theoretical lenses to report these limits. Section 3.3 exposes how AI-driven digital marketing could be a stimulus for the sustainability of the supply chain. Section 3.4 debates the perceptions, obstacles, and ethical considerations connected with these enhancements. Finally, Sect. 3.5 brings a conclusion, reviewing the main results, detecting provisions for theory and practice, and indicating future research guidelines.

3.2 Theoretical Essential and Research Gaps

3.2.1 Recap of Present Theories in Sustainability and Digital Marketing

The appreciation of the theoretical foundations of marketing and consumer behavior is vital for understanding the influence of AI in sustainability encouragement. Different confirmed theories have been utilized traditionally to describe and forecast customer alternatives, including those relevant to sustainable consumption. It furnishes a fundamental lens by which to inspect the composite interactions of psychological, social, and environmental forces affecting customer decisions.

The Theory of Reasoned Action (TRA) (Fishbein and Ajzen 1977) predicts that the behavioral intention of an individual is fixed by their conduct across the behavior and subjective norms. Theory of Planned Behavior (TPB), which is an extension of TRA, combines the recognized behavioral regulator as a third cause, recognizing that entities may not have broad control over their activities. In the sustainable consumer behavior background, the main theory and its extension have been employed to realize aspects inducing intentions to buy eco-friendly products, involved in recycling, or implement sustainable existences. For example, a customer's intent to purchase an organic product could be motivated by their affirmative approach to organic food, their peers' noticeable pressure to buy organic, and their belief in their ability to find and afford such products. While these theories deliver a strong framework for considering individual intents, their claim in the fast-evolving digital scene, mainly with AI-driven interferences, often needs extension and enhancement.

The Social Cognitive Theory (SCT) places a strong emphasis on how social experience, self-efficacy, and observational learning all contribute to personality development. SCT contends that people pick up behaviors from watching others and the results of their actions when it comes to consumer behavior. Self-efficacy, or the conviction that one can carry out the actions required to achieve particular performance goals, is important. SCT can be used in sustainable marketing to explain how consumers follow sustainable practices by looking at what social groups, influencers, and role models do. Due to their ability to foster social sharing and community development, digital platforms offer an ideal environment for the implementation of SCT, as users are continuously exposed to the actions and viewpoints of others. Still more research is needed to fully understand how AI directly affects these social learning processes and the development of self-efficacy in sustainable consumption.

Diffusion of Innovations Theory, developed by Everett Rogers' provides an explanation of how, why, and how quickly new technologies and concepts are adopted by different cultures. Relative advantage, compatibility, and complexity are attributes that affect adoption observability as well as trialability. It has been applied to sustainable marketing. Acquiring the new eco-friendly activities' agreement; alternatives, their ease of use, and their suitability for current lifestyles, will be tested and watched. While useful for understanding the spread of sustainability innovations, this theory predicts that the application might not adequately convey the subtleties. How AI can speed up or improve processes with its personalized and adaptable capabilities, impede the adoption of sustainable practices and technology.

3.2.2 Identifications of Research Gaps

Concerning the expansion of AI-driven digital marketing and sustainability literature, many significant research gaps persist. Highlight spots where existing theoretical frameworks and empirical studies fail in delivering an inclusive understanding of the difficult exchanges. Seizing these gaps is dynamic for evolving both the academic and practical sides in the field. While existing literature acknowledges AIs role in sustainable marketing and consumer behavior there is a limited understanding of the specific mechanisms through which AI influences sustainable consumer choices particularly concerning supply chain sustainability, many studies describe what, e.g., AI can personalize messages but not the how or why these AI interventions lead to sustainable outcomes for instance how does an AI-powered recommendation system precisely alter a consumers decision-making process to favor an eco-friendly product what cognitive or emotional pathways are activated a deeper mechanistic understanding is needed to design more effective and ethical AI interventions (Segun-Ajao 2025). The nexus among AI and ecological social and economic performance, the three theoretical corner-stones of business sustainability has not been inspected to a substantial range exclusively in the digital marketing framework and supply chains whereas a few examine upon individual features, e.g., AI for waste reduction a holistic examination of how AI-driven digital marketing participates to a business's total sustainability functioning around the three dimensions is missing this gap is mainly related for understanding the wider effect of AI on sustainability of the supply chain which fundamentally comprises of ecological social and economic reflections through its value chain (Lopes de Sousa Jabbour et al. 2025).

The expanding consumer interest in ethical and transparent business practices remains poorly understood, including the impact of AI on the pressure for businesses to comply and how transparency through AI drives company responses. Intending to address these questions, it is essential to explore how AI can act as the spine of transparency efforts by successfully communicating the full supply chain story to customers, for instance, in a way that resonates with CPEC's ideologies and shapes trust. What are the inferences of AI-driven transparency for customer insights and consequent behavior? Is there a variance in customer purchase intention when CPEC transparency is presented as an AI-driven transparency vs traditional, text-based forms that can be difficult to read or dull to customers, and may not grasp their attention so fast? Graded as the results may be, though, this focus on a great omission as it neglects the conception of advanced interchange through the way customers claim businesses, the AI that can fulfill such tasks, and how ethics are tailored throughout supply chain functions (Holloway 2024).

3.2.3 Absence of Novel Theoretical Frameworks for Interdisciplinary Analysis

Recent investigations often depend on well-known theories (e.g., "TRA/TPB, SCT, Diffusion of Innovations"), but it is essential to employ innovative theoretical lenses to advance greater perceptions into the complicated interaction of AI-driven digital marketing, sustainable consumer behavior, and the sustainability of the supply chain. The AI's rapid growth and its incorporation into several features of digital marketing and supply chains necessitate novel theoretical perceptions that can gain exclusive dynamics and evolving phenomena. Depending exclusively on usual theories may restrict the investigation range and delay the growth of advanced resolutions.

Outline of the Novel Theoretical Lenses. To investigate the recognized gaps and furnish a deeper appreciation of the convergence between AI-driven digital marketing, sustainable consumer behavior, and the sustainability of the supply chain, this chapter suggests the implementation of distinct advanced theoretical lenses. These theories, derived from several disciplines such as sociology and organizational studies, provide novel viewpoints and analytical tools to examine the composite technique and dynamics at play.

Nudge Theory (Behavioral Economics) for AI-Driven Involvements. Nudge Theory, derived from behavioral economics, proposes that indirect interferences can significantly affect the choices and behaviors of individuals with no limitations on the alternatives or crucially changeable financial incentives. The nudge is any feature of the alternative structure that converts the behavior of individuals without preventing any choices or seriously modifying their financial incentives. To incorporate it as only a nudge, the involvement must be simple and affordable to bypass. Nudges are not directive. In the era of AI-driven involvement, Nudge Theory can be utilized to recognize how AI can indirectly affect customer options across sustainable products and executions, mainly when the information transparency of the supply chain is offered. For instance, an AI-driven online store could indirectly highlight eco-friendly options, renewed search outcomes to prefer sustainable alternatives. This theory enables describing the AI-driven nudges outline and validation in marketing eco-friendly buying decisions without compulsory action adoption.

Extended Theory of Planned Behavior (TPB) with AI-Mediated Influences. Even though the traditional Theory of Planned Behavior (TPB) gives a good basis for analyzing behavioral intentions, an adapted TPB would address how AI influences consumer behavior. This extension would examine the attitudes of consumers toward social norms and their perceived eco-friendly behavior regarding AI-driven information, tailored suggestions, and interactive features, as well as their perception of social norms' influence on performing sustainable actions. For example, AI can offer personalized information highlighting the environmental cost of certain products, illustrate peer sustainable consumption as social proof, or provide virtual fittings that bolster a consumer's confidence in making sustainable purchases. Understanding the psychological processes of AI-initiated sustainable behaviors, especially when detailed information on the sustainability of the supply chains is provided by AI, would be aided by this extended framework (Dwivedi et al. 2021).

Social Practice Theory for AI-Transformed Consumption and Supply Chain Practices. Social Practice Theory (SPT) provides one of the best paths to explaining human actions by viewing them as interrelated and not separated activities (Lu et al. 2016). SPT covers the components that establish the action, which are meanings, capabilities, and materials. In the Sustainability environment, SPT can investigate the ways by which AI changes current consumption and supply chain activities to more sustainable ones. In procurement, AI can automate the process of picking out sustainable suppliers and simplify the goods transportation to decrease emissions, which changes the logistics from carbon-heavy to sustainable ones (Wang et al. 2020). This theory offers the systemic AI effects on daily life routine as well as its organizational incorporation and moves actions further on individual decisions.

Institutional Theory for Organizational Adoption of AI for Sustainability. Concerning organizations, Institutional Theory emphasizes how corporations implement precise activities and structures to maximize effectiveness and acquire legitimacy for investors and adapt to societal expectations. It emphasizes institutional forces of coercive, derivative, and regulating effects on organizational behavior. Concerning AI-driven sustainability, the theory can study how exogenous aspects such as market forecasting, policies, or industry criteria affect the AI-driven sustainability in marketing and the management of the supply chain. For instance, increasing normative consumer demand for transparent supply chains may motivate firms to adopt and invest in AI traceability solutions. Likewise, new environmental regulations would likely force firms (often called 'coercive forces') to adopt AI for resource consumption optimization. This theory sheds light on the rational motives of leveraging AI for sustaining business initiatives irrespective of the immediate economic gains, as it enables organizations to retain credibility and stave off penalties.

3.3 Supply Chain Sustainability Pathways Through AI-Driven Digital Marketing

AI and digital marketing integration is restructuring consumer behavior and the functioning spine of supply chains, providing transformative sustainability paths. By enhancing procedures from sourcing to distribution, AI can successfully decrease the ecological effect, improve the social responsibility, and expand economic productivity within supply chain actions.

3.3.1 AI for Supply Chain Transparency and Traceability

Supply chain transparency (SCTrans) is a tactical necessity as shareholders claim evidence of sustainability and ethical behavior. AI is dominant in getting the transparency, empowering solid data gathering, confirmation, and interaction over the supply chain.

AI Utilization for Data Gathering and Validation. Institutional theory declared that in a way to reveal transparency, firms face growing pressures, such as regulatory requests and customer anticipations. The systems of AI can gather and check a great bulk of data from each point within the supply chain, encompassing ecological, social, and economic metrics. It can also confirm data validity by exposing conflicts that may present unsustainable actions. For instance, AI can utilize blockchain to generate fixed documents for the origin of the product, improving trust and traceability. It can systematize deference supervision by constantly examining supplier commitment to ecological and social norms, decreasing physical audit loads, and increasing accuracy. This transparency for the data-driven approach allows businesses to form perfect sustainability outlines, which can be transmitted to shareholders and applied to ESG documents.

Adopting Customer Pressure for Ethical Conduct (CPEC) Through Transparency. Customer Pressure for Ethical Conduct (CPEC) is an influential stimulus for supply chain conversion. AI-powered digital marketing can provide shared product information. AI allows customers to retrieve comprehensive supply chain data through QR codes or digital platforms, exposing source, production environments, and accreditations in a simply understandable presentation. Create tailored sustainability reports, AI can offer customers personalized data within the ecological and social influence of their acquisitions, and nurture loyalty. Distribute immediate information, AI can inform shareholders of the main supply chain actions, such as waste decrease accomplishments, transparency support, and assigned accountability.

Preventively, revealing translucent information, firms can adapt customer load into a substance for constructive supply chain modification, construction competitive advantage, and trust.

3.3.2 AI-Guided Resource Optimization and Wastage Reduction

AI forecasting and statistical abilities are playing a critical role in improving resource effectiveness while diminishing waste alongside the supply chain. AI has a crucial role in demand prediction and inventory controlling; it measures data that is historical in nature, market trends, and other aspects (such as social media) to offer appropriate demand estimates, thus avoiding unnecessary production along with cost. AI-guided inventory technologies expand inventory management in storage facilities, declining overload warehousing and the opportunity of expiration. This advocates the operative application of important resources, thereby considerably dropping the ecological footprint. AI executes an important function in innovative packaging and monitoring the goods lifecycle; it might suggest fit shield in measuring its structures and paths for delivery, declining used materials, and recyclable progressing or ecological choices. AI improves product lifecycle management (PLM) by estimating the desired maintenance, simplifying modifications, and linking customers with green challenges, consequently furthering concepts of the circular economy and declining the wastage. AI has an integral part in improving energy effectiveness and enabling the inclusion of green power, consequently minimizing the power consumption impact of processes within the supply chain. AI is necessary for managing the data center and warehousing; it may constantly adjust system deliveries, cooling down, and power in data centers, along with HVAC and illumination in storefronts, depending on actual demand and the surrounding conditions, thus decreasing wasted energy.

Sponsoring Ecological Logistics. AI-driven logistics might merge shipments, route improvement, and join ecological transport, thus declining power usage and emissions. Actual traffic and weather investigation simplifies active switching, henceforth improving logistical productivity and ecological functioning.

3.4 Insights and Challenges: Ethical, Social, and Managerial Inferences

Concerning the inclusive possibility of AI-driven digital marketing to advance the sustainability of customer behavior and supply chains, its employment meets some obstacles. An important estimation of concepts and difficulties is critical for the effective employment of AI-powered sustainability.

3.4.1 Ethical Considerations of AI-Driven Sustainable Digital Marketing

The AI ethical measures are essential, particularly when dealing with critical contexts such as customer behavior and ecological effects. Assure that AI systems are produced and exploited responsibly is fundamental to boost trust and attain authentic, sustainable outcomes.

Privacy and Data Security Concerns. Collecting, analyzing, and explaining customer data are the pillars of AI-powered sustainable marketing. This advances substantial data privacy security disputes. Customers may be worried about sharing individual information, even if it is for an apparently philanthropic reason, such as sustainability encouragement. Organizations must execute strong data shield metrics, stick to authoritarian rules for privacy, and be clear in stating how they will collect and use their data. If they did not do so, it would result in eroding their trust, drive to backlash, and hinder the sustainable marketing goals.

Algorithmic Bias and Fairness in Sustainability Messaging. Genetic data is the basis for AI algorithms, which can accidentally have prejudices that reveal societal disparities. If it is not precisely handled, these biases can be sustained or even boosted by AI systems, leading to inequitable results in sustainability messaging. For instance, an AI might excessively choose specific segments with sustainable product interests and abandon others, or it might unintentionally support stereotypes about who is or is not sustainability-fascinated. To ensure the fairness of the algorithm, various and prototypical training data is needed, consistent tracking for bias, and the execution of moral AI design essentials to certify that sustainability messages are transmitted equitably to all segments.

Greenwashing Risk and Customer Trust. The sustainability appeal might rarely drive firms to involve in "greenwashing", generating ambiguous statements on their ecological routine. AI, if imperfectly structured, could unconsciously simplify greenwashing by creating influential but imprecise ecological-value communications or by enhancing campaigns that encourage uncertain assertions. This pretends a substantial risk to customer trust. To diminish this threat, the models of AI have to include processes to validating claims of the sustainability, comparing data with reliable independent authorizations, and highlighting possibly deceptive content. Claiming accessibility and clarity in sustainable messaging is imperative for construction and sustaining customer trust in AI-powered sustainable marketing programs.

3.4.2 Social Consequences: Customer Acceptance and Behavioral Change

The AI accomplishment in controlling ecological customer behavior relies on customer tolerance and their willingness to accept innovative, more sustainable activities. This includes accepting and adopting possible social inferences.

Overpowering Reluctance to AI-Driven Nudges. When Nudge Theory advises that detailed mediations can be well organized, customers may show reluctance if they understand AI-driven nudges as interfering. Excessively constant references, even if well-intentioned, can drive a surveillance sensation or being controlled.

To cope with this reluctance, AI intrusions should be shaped to be translucent, explainable, and supply customers with a feeling of power and control over their options.

3.4.3 Managerial Challenges and Future Directions

AI implementation for sustainability within the digital marketing and supply chain era raises various managerial challenges that businesses should address to reveal their filled potential.

Data Integration and Interoperability. Operative AI-powered sustainability inventiveness involves the unified incorporation of various datasets from several bases, containing marketing platforms, the system of supply chain management, IoT strategies, and peripheral ecological data. Attaining data harmony across different systems and divisions can be an important practical and organizational provocation. Businesses require to place in solid data infrastructure, set up comprehensive data governance strategies, and boost cross-functional integration to guarantee that AI systems have access to the high-quality and inclusive data they seek to create precise perception and drive sustainable consequences.

Skill Gaps and Organizational Readiness. AI sustainable adoption needs qualified staff in AI enhancement, data science, and sustainable business activities. Yet many organizations have skill gaps in these contexts. Moreover, readiness at the organizational level, such as innovation abilities and verifiable administrative commitment to sustainability, is crucial. Firms should invest in training and reskill programs for their staff, enhance the shaping of interdisciplinary groups, and stimulate a culture that boosts the rooting of AI within sustainability strategies.

Balancing Profitability with Sustainability Goals. Although AI can provide both ecological and economic gains, organizations face the provocation of leveling profitability and sustainability objectives. Starting developments in AI for sustainability can be essential, and the return on investment may not always be instantly visible. Managers seek to improve direct metrics to size the ecological, social, and economic influence of their AI-powered initiatives of sustainability. They must also conduct a visibility study that reveals how these actions add to long-term value creation, fostering the reputation of the brand, and increasing customer loyalty and retention, thereby stratifying profitability with sustainability goals.

3.5 Conclusion and Future Research

3.5.1 Finding Summary

The growing ecological emergency and rising social consciousness have highlighted the significant requirement for sustainable activities in all fields (Zabaniotou 2020). In the commerce era, an important move is proceeding, shifting away from out-of-date profitable options to focus on ecological preservations and ethical behavior. This model modification is mainly obvious in customer behavior, as a greater number of people are looking for goods and services that associate with their sustainable principles (Purcărea et al. 2022). Sustainable customer behavior, including activities such as buying ecological goods, decreasing usage, and appointing in reprocessing, has moved from a minority issue to common practice. The age of digitization, described by extensive interaction and huge change in the information, intensifies both the obstacles and possibilities in enhancing these types of actions (Rachinger et al. 2019).

3.5.2 Contribution to Theory and Practice

The current chapter proposes various important aspects for academic and practical attention. Academically, it expands current frameworks by merging a new point of view to demonstrate the complicated interaction through AI, digital marketing, sustainable consumer behavior, and supply chain sustainability. By employing Nudge Theory, we have clarified how precise AI-powered involvement can significantly affect customer decisions. The TPB extension allows a more inclusive paradigm for conceptualizing AI's role in customer intents across sustainable activities. Social Practice Theory provides a comprehensive point of view of how AI can convert consumption and supply chain activities to more sustainable procedures. Eventually, Institutional Theory assists in explaining the organizational motivations behind the AI adoption or sustainability actions. For the practical aspect, the current chapter delivers a broad blueprint for businesses and decision-makers seeking to utilize AI in digital marketing to boost sustainable customer behavior and shape a stronger, environmentally responsible supply chain. It underlines legal strategies for personalized green messaging, recommendations for sustainable products, and optimization of the supply chain; however, it also affirms the crucial significance of ethical reflections.

3.5.3 Limitations and Future Research

Although comprehensive of the current chapter, it admits standing limitations that stock scope for more research. In addition to suggesting new conceptual implementations, we need empirical evidence of this structure in the AI-powered scenes of sustainable marketing. Upcoming studies could perform quantitative studies to check the hypotheses developed from these theories and then expand empirical proof for the efficiency of AI inferences. Furthermore, AI ethics should hold on to be examined, with a specific coverage on such issues as privacy of data, and biases of algorithms. Longitudinal studies should be employed on the AI's long-term influences on consumer behavior in addition to supply chain actions, moving away from simply short-term behavior modifications to long-term shifts of sustainability. One more crucial issue for further studies is the function of regulatory frameworks and global alliances in AI governance for sustainability. And ultimately, besides the growth of AI technologies, further research must estimate the probability that emerging AI paradigms (for example, explainable AI) can propose to stimulate sustainable customer behavior and supply chain sustainability to align technological progression with ecological and social instructions.

References

Dwivedi YK, Ismagilova E, Hughes DL, Carlson J, Filieri R, Jacobson J, Jain V, Karjaluoto H, Kefi H, Krishen AS, Kumar V, Rahman MM, Raman R, Rauschnabel PA, Rowley J, Salo J, Tran GA, Wang Y (2021) Setting the future of digital and social media marketing research: perspectives and research propositions. Int J Inf Manag 59. https://doi.org/10.1016/j.ijinfomgt.2020.102168

Fishbein M, Ajzen I (1977) Belief, attitude, intention, and behavior: an introduction to theory and research

Holloway S (2024) Unveiling customer perceptions: a qualitative study on the role of supply chain transparency in brand trust. Preprints 1–10. https://doi.org/10.20944/preprints202406.1738.v1

Lopes de Sousa Jabbour AB, Laguir I, Stekelorum R, Gupta S (2025) The nexus of artificial intelligence and sustainability performance: unveiling the impact of supply chain transparency and customer pressure on ethical conduct. J Environ Manage 379:124847. https://doi.org/10.1016/J.JENVMAN.2025.124847

Lu B, Fan W, Zhou M (2016) Social presence, trust, and social commerce purchase intention: an empirical research. Comput Hum Behav 56. https://doi.org/10.1016/j.chb.2015.11.057

Mossavar-Rahmani F, Zohuri B (2024) Artificial intelligence at work: transforming industries and redefining the workforce landscape. J Econ Manag Res 1–4. https://doi.org/10.47363/JESMR/2024(5)213

Pappas IO, Mikalef P, Dwivedi YK, Jaccheri L, Krogstie J (2023) Responsible digital transformation for a sustainable society. Inf Syst Front 25(3):945–953. https://doi.org/10.1007/s10796-023-10406-5

Purcărea T, Ioan-Franc V, Ionescu ŞA, Purcărea IM, Purcărea VL, Purcărea I, Mateescu-Soare MC, Platon OE, Orzan AO (2022) Major shifts in sustainable consumer behavior in Romania and retailers' priorities in agilely adapting to it. Sustainability (Switzerland) 14(3):1–51. https://doi.org/10.3390/su14031627

Rachinger M, Rauter R, Müller C, Vorraber W, Schirgi E (2019) Digitalization and its influence on business model innovation. J Manuf Technol Manag 30(8):1143–1160. https://doi.org/10.1108/JMTM-01-2018-0020

Rasheed A, San O, Kvamsdal T (2020) Digital twin: values, challenges and enablers from a modeling perspective. IEEE Access 8. https://doi.org/10.1109/ACCESS.2020.2970143

Ruswandi I, Khaeru Ahmadi RA (2023) Maximize the use of internet to improve sales of small businesses: evidence from Indonesia. Tech Soc Sci J 50. https://doi.org/10.47577/tssj.v50i1.9887

Segun-Ajao E (2025) AI and sustainable procurement: a path to green supply chains, pp 1–16. https://doi.org/10.20944/preprints202501.1516.v1

Theodorakopoulos L, Theodoropoulou A (2024) Leveraging big data analytics for understanding consumer behavior in digital marketing: a systematic review. Hum Behav Emerg Technol 2024. https://doi.org/10.1155/2024/3641502

Wang K, Zipperle M, Becherer M, Gottwalt F, Zhang Y (2020) An AI-based automated continuous compliance awareness framework (CoCAF) for procurement auditing. Big Data Cogn Comput 4(3). https://doi.org/10.3390/bdcc4030023

Yadav R, Pathak GS (2017) Determinants of consumers' green purchase behavior in a developing nation: applying and extending the theory of planned behavior. Ecol Econ 134. https://doi.org/10.1016/j.ecolecon.2016.12.019

Zabaniotou A (2020) A systemic approach to resilience and ecological sustainability during the COVID-19 pandemic: human, societal, and ecological health as a system-wide emergent property in the Anthropocene. Glob Transit 2:116–126. https://doi.org/10.1016/J.GLT.2020.06.002
Ziakis C, Vlachopoulou M (2023) Artificial intelligence in digital marketing: insights from a comprehensive review. Information (Switzerland) 14(12). https://doi.org/10.3390/info14120664

Artificial Intelligence and the Green Dilemma: Harnessing Innovation for Environmental Sustainability

4

Raed Abdelhaq, Bahaa Awwad, and Ammar Zakaria Salem

Abstract

This study explores the two-faced relationship between artificial intelligence (AI) and environmental sustainability, a reality commonly known as the "Green Dilemma." It explains the manner in which state-of-the-art AI technologies provide innovative solutions to urgent environmental issues such as climate change, resource exploitation, waste management, and biodiversity conservation. It also recognizes the contradictory environmental impact of AI, which is tantamount to enormous energy and resource utilization. Based on recent scholarly literature, the study provides an overview of the application of artificial intelligence in environmental monitoring, renewable-energy systems, circular economy initiatives, and ecosystem management. The description emphasizes the issue of "Green AI," suggesting effective strategies to reduce the adverse environmental consequences of AI through sustainable technological innovation, policy flexibility, and interdisciplinary collaboration. The final study provides policy recommendations and future-oriented suggestions to guarantee AI as an effective force toward environmental sustainability and reduce its ecological liabilities.

R. Abdelhaq
Accounting Department, Faculty of Business and Communications, An-Najah National University, Nablus, Palestine
e-mail: raed.abdelhaq@najah.edu

B. Awwad (✉)
College of Business and Finance, Ahlia University, Manama, Bahrain
e-mail: bawwad@ahlia.edu.bh

A. Z. Salem
Department of Administrative and Financial Sciences , Palestine Technical University—Kadoorie, Tulkarm, Palestine
e-mail: ammar.salem@ptuk.edu.ps

B. Awwad and R. A. Awwad (eds.), *Technology and Innovation Towards a Future of Business Sustainability*, Contributions to Environmental Sciences & Innovative Business Technology, https://doi.org/10.1007/978-981-95-6469-9_4

Keywords

Artificial intelligence • Environmental sustainability • Green AI

4.1 Introduction

4.1.1 Background: The Growing Environmental Crisis and the Role of Technology

The recent acceleration of the global environmental crisis, as exemplified in the increasing rate of climate change, natural resource exhaustion, and biological diversity loss, imperils not only human survival as a species but also the Earth's ecosystem. This calls for new visions and radical human–environment relations transformation in an attempt to tackle this crisis. Technological innovation is becoming a top priority in the context of promoting environmental sustainability (Arthington et al. 2023; Razia et al. 2024; Awwad 2022, 2025). Technologies such as renewable-energy systems and advanced monitoring systems are only beginning to develop competencies to act against environmental degradation and facilitate greener behaviors. Yet, such technologies must be designed and implemented with caution since poorly conceived implementations would unknowingly work toward exacerbating the environmental issues (Dwivedi et al. 2021; Awwad 2024).

4.1.2 Artificial Intelligence: A Force for Sustainability

Artificial Intelligence (AI) technology, since it can handle large volumes of data, identify complex patterns, and improve complex systems, is a paradigm-shifting technology with unrivaled potential to engage with environmental issues. Implementation of AI improves environmental monitoring, resource efficiency, energy use, and conservation in the industries. For instance, AI systems scrutinize satellite images to track deforestation, forecast extreme weather, or model smart-grid layouts for easier integration of renewables (Abdeljawad et al. 2025; Hamdan et al. 2021; Razia and Awwad 2022). The ability of AI to learn and adapt makes it an exciting collaborator on the path to a sustainable future.

4.1.3 Study Goals and Organization

This study critically examines the role played by artificial intelligence (AI) in environmental sustainability. It is meant to map out the wide range of AI-facilitated resources that assist in maintaining the natural world and saving resources, as well as questioning the ecological costs of producing and deploying AI technologies. In this manner, the study does the following:

- Conducts a literature-based, systematic review of the literature on the application of AI in environmental stewardship with an upward slant in the articles in high-impact journals.
- Clarifies the potential benefits and the related issues that are associated with the integration of AI into environmental governance.
- Takes the idea of the Green AI into account and outlines the possible routes to reducing the ecological footprint of AI-reliant systems.
- Thinks over the policy implications and the future of shaping AI to lead to sustainable results.

The study is organized as follows: Sect. 4.2, an elaborate literature review, conducts a survey of the AI-related environmental impact in various areas of application and identifies the related challenges. Section 4.3 looks at methodologies and theories of evaluating and strengthening the environmental performance of AI. Section 4.4 offers a discussion of the two-fold alignment of AI with sustainability, in terms of policy and future research. Section 4.5 sums up the volume with the distillation of key findings and the recommendations for a sustainable path of AI development.

4.2 Literature Review: AI's Impact on Environmental Sustainability

Artificial intelligence (AI) has taken a significant role in modern environmental debate, being generally viewed as a revolutionary solution to complicated social problems. The ability to analyze large amounts of data, identify complex patterns, and optimize system performance provides new solutions in various areas of environmental sustainability (Vinuesa et al. 2020; Awwad (2025a). The ensuing discussion takes a survey of the main AI applications in this regard, outlining the associated benefits and constraints.

4.2.1 AI for Environmental Monitoring and Data Analysis

Environmental monitoring and interpretation today are, to a very large measure, dependent on AI technologies that improve the granularity and dynamism of the ecological knowledge and therefore the efficiency of management. AI systems can be utilized to monitor environmental change, predict events, and control pollution by combining real-time sensor, satellite, and ancillary observation data. For instance, collaboration between machine-learning modules is being used more frequently to query satellite data in an attempt to monitor deforestation, determine land-use change, and place a value on ecosystem health (Alotaibi and Nassif 2024;

Alkababji and Awwad 2023). Such functions allow for the anticipation of degradation and intervention before it happens. Additionally, AI is being used to actively refine intricate climate models so that weather forecasting and climate prediction improve, which is crucial to adaptation planning and disaster planning (Abdelhaq et al. 2025; Awwad et al. 2024). Introduction of AI into environmental monitoring would therefore provide an enhanced, real-time description of ecological conditions that would be out of the realm of the conventional, typically fixed, methods of evaluation.

AI in Renewable Energy and Energy Efficiency AI is a critical tool in the process of optimizing renewable-energy systems and enhancing energy efficiency, in general. In the context of renewable-energy use, AI algorithms can predict the uncertainty of demand and supply and optimize the positioning and efficiency of wind turbines and solar panels, as well as control smart grids to enable the efficient integration of various energy sources (Salem and Awwad 2025). Optimization is translated into efficiency, minimization of wastage, and improved reliability of renewable-energy systems. As an example, predictive maintenance based on AI can identify possible failures in wind turbines before they happen, reduce downtime, and optimize the efficiency of energy production. AI also allows buildings and industrial processes to achieve higher energy efficiency, since it allows tracking the consumption trends and optimizing systems, like heating, ventilation, and air conditioning (HVAC), to reduce energy loss (Amasyali and El-Gohary 2018). AI in such areas is essential to the shift to a low-carbon economy and the achievement of high climate goals.

4.2.2 AI in Circular Economy and Waste Management

The use of a circular economy and waste management is critical in reducing the environmental footprint and ensuring sustainable use of resources. Artificial intelligence (AI) brings new solutions to both fields, especially by optimizing the routes of waste collection and increasing the level of sorting and recycling. Computer-vision systems built on the power of AI show great precision in sorting and classifying various types of waste, which enhances the capacity of recycling centers (Lakhouit 2025). In its turn, predictive analytics will be able to optimize the best collection routes, reducing fuel costs and operating expenses. At a more comprehensive scale, AI is compatible with the circular economy since it promotes the design of products to be durable, repairable, and recyclable, and enables sharing, reuse, and repurposing through digital platforms (Arthington et al. 2023). Finally, through the increased use of resources and a reduction in waste production, AI also plays a significant role in material loop closure and the development of a more sustainable consumption-production system.

4.2.3 AI in Biodiversity Conservation and Ecosystem Management

Ecosystem degradation and loss of biodiversity are two of the most pressing issues for the health of the planet and human health. Artificial intelligence (AI) is currently providing conservation practitioners with data-driven tools that are powerful and effective in the systematic monitoring, analysis, and management of natural resources. By virtue of being able to analyze large amounts of ecological data (acoustic sensors, camera-trap images, drone-mounted images, etc.), AI-driven systems can identify the presence of species, wildlife population maps, and even illegal practices (poaching or unauthorized logging), among others (Rathi 2025). Specifically, the deep-learning algorithms have the potential to analyze audio information to detect the existence of particular animal species, hence offering scalable and non-invasive monitoring capabilities. As part of the wider ecosystem management field, AI can streamline the conservation resource allocation, predict the proliferation of invasive species, and simulate the impact of climate change on ecosystems, and thus guide the practice of adaptive management. These apps can increase the accuracy and the success of conservation projects, which can help to save endangered species and rebuild their necessary habitats.

4.2.4 Challenges and Limitations of AI in Environmental Sustainability

While AI has immense potential in the context of environmental sustainability, its utilization is also paired with a chain of challenges and shortcomings. The most basic and preliminary one is the environmental cost of AI systems themselves: it requires high computing capacity to train and run complex structures, especially deep-learning configurations and large language models, which lead to high energy consumption and growing carbon footprints (Bolón-Canedo et al. 2024). Therefore, data centers, being the physical locations where the servers required to perform AI operations are located, become the largest electricity and water consumers, invoking questions about their role in climate change and resource depletion (Kirikkaleli et al. 2025).

The abrupt release and distribution of extremely powerful generative AI models increase these issues as they are accompanied by a proportional rise in electricity consumption and water absorption (Năstasă et al. 2024; Khoffash and Subhi Awwad 2024).

Beyond explicit environmental limitations, other limitations include the demands for large amounts of good-quality data that might be hard to find and organize in environmental contexts. Ethical considerations such as algorithmic bias and data protection also need to be addressed carefully so AI solutions are equitable and do not inadvertently cause harm to already-vulnerable groups and ecosystems. Moreover, the sheer sophistication of environmental systems typically renders it unrealistic to develop AI models that accurately capture all relevant variables and interactions, threatening to oversimplify or be misleading. To overcome these challenges, it will require a long period of endeavoring to make AI construction more energy-efficient, instating sound regulatory frameworks, and increasing inter-disciplinary cooperation in the hope of making AI an actually sustainable environmental protection technology.

4.3 Theoretical Framework

Artificial intelligence systems have the potential to enhance the work of environmental sustainability as much as they have the potential to bring ecologically harmful implications. There is thus a requirement to construct a serious theoretical framework and supplement empirical methods. These are the guidelines upon which the estimation of the environmental impact of AI is carried out and according to which sustainable AI systems are developed. This discussion addresses two critical topics: (1) lifecycle analysis of AI systems and (2) the new field of Green AI (Schwartz et al. 2020).

4.3.1 Measurement Frameworks for the Environmental Footprint of AI

The overall vision of AI systems' environmental impact requires an integrated approach that breaks down the whole technological lifecycle, starting with the hardware production, then data-center use, and ending with the energy consumption at model training and inference. The effect may be quantified in relation to conventional environmental measurement methods, with Life Cycle Assessment (LCA) among them being widely prominent. For AI, an LCA would include the material and energy inputs used to manufacture AI hardware devices like GPUs and TPUs, and the water and electricity used in computation and cooling in data centers. Alternatively, it would also be possible to have a parallel structure that would include the energy used to transmit and store data. This kind of holistic LCA offers a full picture of the environmental footprint of AI, thus enabling us to pinpoint its hotspots and strategize on which ones to prioritize for mitigation. Although a standardized LCA of AI remains in the making, adoption of a whole-of-environment accounting approach is still essential in order to make the transparent and accountable practice of AI development (Strubell et al. 2019).

4.3.2 Green AI: Methods of Sustainable AI Design

The emergence of the so-called Green AI in the last few years has been driven by the mounting environmental issues related to the advancement of more efficient computing methods. Green AI means a collection of practices and procedures that are especially utilized to reduce the energy requirements and their related carbon footprint of AI systems throughout the cradle-to-grave lifecycle of AI systems (Bolón-Canedo et al. 2024; Schwartz et al. 2020). An end-to-end green AI paradigm thus comprises a number of self-sustaining characteristics. All these dimensions, which are all consistent with the different technical, economic, environmental, and social factors, are compatible with one another. The structure can be summed up as follows:

- Energy-efficient algorithms: The challenge of creating and implementing machine-learning models that require less computational power during training and inference can be solved in three general strategic directions: optimization of model architecture, efficient training strategies, and post-training (after training) methods of model compression and quantization (Patterson et al. 2021). State-of-the-art advances in each of the three domains have the promise of significantly reducing the computational overhead of the current generation of deep-learning systems.
- Hardware optimization: The efficient design and implementation of hardware systems that take advantage of low power consumption and are specifically and uniquely optimized to run artificial intelligence (AI) workloads represent an important research field. In this field, research has resulted in the improvement of chip architectures and the creation of special AI accelerators that work with less power consumption (Abdelhaq et al. 2024).
- Sustainable data management: Sustainable data management requires the use of practices that reduce the amount of energy used in storing, processing, and transmitting data. These practices include data pipeline optimization, the deployment of storage devices that consume less power, and avoiding the generation of redundant data sets (Kommineni and Chundru 2025).
- Moral deployment: The implementation of artificial intelligence (AI) systems in environments with renewable-energy sources and equipped with an energy-efficient cooling architecture is a significant technological challenge. What is especially worrying is that the model has to be retrained regularly to reduce the amount of energy spent continuously.

The adoption of the values of Green AI by scientific and practitioner communities is a decisive moment when the transformative potential of artificial intelligence (AI) is used to develop the planet in a sustainable way and reduce the negative impact on the natural environment to a minimum. The main idea of this paradigm is to shift the direction of AI development toward an ethical and ecological responsibility, as shown in Fig. 4.1.

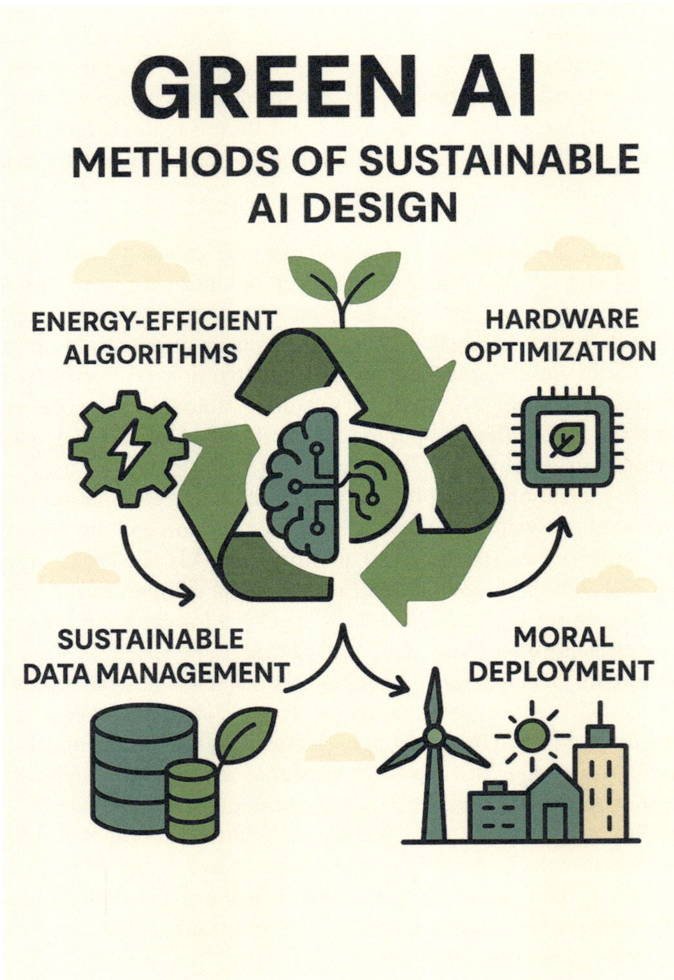

Fig. 4.1 Green AI: methods of sustainable artificial intelligence design. *Source* Prepared by the authors

4.4 Policy Implications and Institutional Frameworks

To ensure that the AI production is in line with environmental sustainability objectives, there should be strong policy implications and regulatory policies. The world agencies and governments play a key role in facilitating the production of Green AI and prohibiting activities that are environmentally unfriendly. Policies can be:

- **Energy efficiency standards**: Establishing energy efficiency standards for AI data centers and hardware to promote the use of renewable sources of energy for AI (Schwartz et al. 2020).
- **Disclosure and reporting**: Requiring disclosure and reporting of AI hardware and model energy usage and carbon emissions by AI deployers and developers (Lacoste et al. 2019).
- **Research grants**: Rewarding research in energy-efficient AI algorithms, sustainable AI hardware design, and environmental cost measuring metrics for AI (Kommineni and Chundru 2025).
- **Green AI incentives**: Policy incentives and economic incentives for AI deployment with significant effects on environmental sustainability, e.g., renewable energy, circular economy, and conservation.
- **Ethical principles**: Implementing ethical principles that integrate environmental concerns in AI development, thus ensuring not only that AI solutions work but also that they are ecologically sustainable (Dwivedi et al. 2021). These can create a wholesome environment for scientific AI, which encourages innovation with minimal negative effect on the environment.

4.4.1 Future Directions for Research and Implementation

All future AI for sustainable environments are dependent on continued research and new methods for implementation. Some of the areas to be addressed in the future include:

- Making AI more energy efficient: A long-term research agenda priority for new algorithms, hardware, and software optimizations that reduce the need for energy without sacrificing performance (Patterson et al. 2021).
- Standardization of environmental footprint: Creating widely adopted standards for measurement and reporting the environmental footprint of AI systems such that comparison and accountability are easy (Strubell et al. 2019).
- Cross-disciplinary coordination: Encouraging more intense coordination between AI researchers, environmental scientists, policymakers, and industry actors in co-designing environmentally friendly but technologically suitable technologies (Vinuesa et al. 2020).
- Climate change adaptation and resilience AI: Improving the use of AI in climate change adaptation and resilience, e.g., predicting early warning systems for extreme climate, resilient agriculture, and infrastructure planning to be resilient (Awwad 2025b).
- Covering the entire lifecycle: From the energy consumption in operations to considering the environmental impact of the entire AI lifecycle, from resource extraction to the disposal of hardware waste at the end of life (Kommineni and Chundru 2025).

Given priority, these issues will enable the AI community to work toward a future in which AI is not only a force for advancement but a foundation upon which a truly sustainable society is established.

4.5 Conclusion

4.5.1 Summary of Key Findings

The complex and dynamic dependency between Environmental Sustainability and Artificial Intelligence (AI) is mapped in this study. The study focuses on the massive potential of AI to serve environmental goals, quoting the obvious utilization of AI in environmental surveillance, renewable-energy optimization, managing waste, and conserving biodiversity. AI gives the power of pattern recognition and optimization of systems by processing enormous and intricate sets of data, thus giving the unprecedented chance of solving the worst environmental problems. The study, however, does not merely critique the overwhelming environmental expense of AI, which is the gigantic energy and water consumption in training and running large models. This seeming contradiction between the two is meant to draw attention to the necessity of an equilibrium viewpoint that can take into account not only the revolutionary potential of AI but also its ecological cost, which it inevitably imposes.

4.5.2 Suggestions Toward a Sustainable AI Future

As a measure to optimize the advantages of AI for environmental sustainability and minimize its drawbacks, some of the strongest recommendations emerge:

- First Priority Green AI Research and Development: Investment in research as a priority in designing energy-efficient hardware, algorithms, and green data policies. This encompasses research into new AI architecture and optimization techniques that reduce the computational need (Schwartz et al. 2020).
- Conduct Complete Environmental Impact Assays: Establish and adhere to best practice in quantifying and reporting the environmental impact of AI systems across their lifecycle. Transparency of this kind is essential to accountability and to truly well-informed decision-making (Strubell et al. 2019).
- Encourage Interdisciplinary Collaboration: Encourage collaboration among AI scientists, environmental researchers, policymakers, and business leaders. This will determine the development of technologically advanced and environmentally friendly AI solutions (Vinuesa et al. 2020).
- Design Supportive Policy and Regulatory Environments: Governments and international bodies should put policies in place that will ensure sustainable AI development, have obligatory energy efficiency policies, and encourage

planning for responsible deployment, like green AI investment and ethics requirements, like consideration of the environment (Dwivedi et al. 2021).

- Promote Education and Awareness: Promote education and awareness among the public, developers, and users about the environmental impact of AI and the value of sustainable AI practices (Aggarwal 2023).

4.5.3 Final Thoughts

Artificial Intelligence (AI) is standing at a fork in its path of development with immense possibilities to benefit the world toward ecological sustainability. In order for this to happen, though, there is a requirement for purposeful and collective efforts to minimize the environmental impact of AI systems. By aligning with the ethical principles of Green AI, developing effective regulations, and fostering intersectoral collaborative innovation, AI can be reversed from becoming a burden to the environment to an active force to sustain goals. An upright initiative toward AI development is imperative in establishing a greener, more sustainable world (Razia et al. 2022).

References

Abdelhaq R, Dwekat A, Atout S, Nour AI (2024) The impact of board characteristics on the level of voluntary disclosure: evidence from Palestinian listed companies. In: Artificial intelligence and economic sustainability in the era of industrial revolution 5.0. Springer Nature Switzerland, Cham, pp 107–122

Abdelhaq R, Salem A, Rabaia DJ, Jardaneh L, Nour AI, Al-Sartawi AMM (2025) Corporate governance and intellectual capital efficiency: empirical evidence from Palestine. Int J Intellect Prop Manag 15(2):180–203

Abdeljawad I, Rabaia DJ, Abdelhaq R, Salem A (2025) The association between corporate social responsibility disclosure and dividend policy: insights from an emerging economy. Int J Disclosure Gov 1–19

Alkababji MW, Awwad BS (2023) Big data tools and techniques in developing accounting practices and the impact of their analysis on the financial performance. Cent Eur Manag J 31(1):618–631

Alotaibi E, Nassif N (2024) Artificial intelligence in environmental monitoring: in-depth analysis. Discov Artif Intell 4(1):84

Amasyali K, El-Gohary NM (2018) A review of data-driven building energy consumption prediction studies. Renew Sustain Energy Rev 81:1192–1205. https://doi.org/10.1016/j.rser.2017.04.095

Arthington AH, Tickner D, McClain ME, Acreman MC, Anderson EP, Babu S, Yarnell SM (2023) Accelerating environmental flow implementation to bend the curve of global freshwater biodiversity loss. Environ Rev 32(3):387–413

Awwad BS (2022) The role of environmental disclosure in enhancing the market value of industrial and service companies listed on the Palestinian stock exchange. Int J Entrep 26(2):1–12. ISSN: 1939-4675

Awwad B (ed) (2024) The AI revolution: driving business innovation and research: volume 1, vol 524. Springer Nature

Awwad B (2025a) Climate finance and sustainability: a story told through the journal of cleaner production. EuroMid J Bus Tech-Innov 31–47

Awwad BS, Alkababji MW, Razia BS (2024) Artificial intelligence and the quality of accounting information in Palestinian industrial companies. J Financ Rep Account (ahead-of-print)

Awwad BS (ed) (2025b) Sustainability in light of governance and artificial intelligence applications. Emerald Group Publishing

Awwad BSA (ed) (2025) AI and IoT: driving business success and sustainability in the digital age. Springer, Cham. https://doi.org/10.1007/978-3-031-92240-4

Bolón-Canedo V, Morán-Fernández L, Cancela B, Alonso-Betanzos A (2024) A review of green artificial intelligence: towards a more sustainable future. Neurocomputing 599:128096. https://doi.org/10.1016/j.neucom.2024.128096

Dwivedi YK, Hughes L, Baabdullah AM, Ribeiro-Navarrete S, Giannakis M, Al-Debei MM, Williams MD (2021) Artificial intelligence (AI): multidisciplinary perspectives on emerging challenges, opportunities, and agenda for research, practice and policy. Int J Inf Manag 57:101994. https://doi.org/10.1016/j.ijinfomgt.2019.08.002

Hamdan A, Hassanien AE, Khamis R, Alareeni B, Razzaque A, Awwad B (eds) (2021) Applications of artificial intelligence in business, education and healthcare, vol 954. Springer, Cham

Khoffash S, Subhi Awwad B (2024) The challenges and opportunities of green financing in Palestine from the perspective of banks. J Infrastruct Policy Dev 08(02):1–20. https://doi.org/10.24294/jipd.v8i12.8800

Kirikkaleli D, Ali K, Zhang Q, Ozrecberoglu Kirikkaleli N (2025) Environmental sustainability in the USA: role of artificial intelligence. Sustain Futures 9:100823. https://doi.org/10.1016/j.sftr.2025.100823

Kommineni M, Chundru S (2025) Sustainable data governance implementing energy-efficient data lifecycle management in enterprise systems. In: Driving business success through eco-friendly strategies. IGI Global Scientific Publishing, pp 397–418

Lacoste A, Luccioni A, Schmidt V, Dandres T (2019) Quantifying the carbon emissions of machine learning. arXiv preprint arXiv:1910.09700

Lakhouit A (2025) Revolutionizing urban solid waste management with AI and IoT: a review of smart solutions for waste collection, sorting, and recycling. Results Eng 104018

Năstasă A, Dumitra T-C, Grigorescu A (2024) Artificial intelligence and sustainable development during the pandemic: an overview of the scientific debates. Heliyon 10(9):e30412. https://doi.org/10.1016/j.heliyon.2024.e30412

Patterson D, Gonzalez J, Le Q, Liang C, Munguia LM, Rothchild D, Dean J (2021) Carbon emissions and large neural network training. arXiv preprint arXiv:2104.10350

Rathi P (2025) Navigating the landscape: challenges and opportunities of AI in wildlife conservation. In: AI and machine learning techniques for wildlife conservation, pp 159–182

Razia B, Awwad B (2022) A comprehensive review of blockchain technology and its related aspects in higher education. In: Technologies, artificial intelligence and the future of learning post-COVID-19: the crucial role of international accreditation, pp 553–571

Razia B, Awwad B, Taqi N (2022) The relationship between artificial intelligence (AI) and its aspects in higher education. Dev Learn Organ Int J 37(3):21–23

Razia AS, Alkababji MW, Ruzieh AS, Awwad BS, Razia BS (2024) Competitive advantage in light of AI to achieve innovative as a key of the SDGS: Palestinian insurance industry. J Lifestyle SDGs Rev 4(1):e01870

Salem AZ, Awwad BS (2025) Does competitive intelligence enhance the market position of Palestinian companies? In: Sustainability in light of governance and artificial intelligence applications, p 209

Schwartz R, Dodge J, Smith NA, Etzioni O (2020) Green AI. Commun ACM 63(12):54–63. https://doi.org/10.1145/3381831

Strubell E, Ganesh A, McCallum A (2019) Energy and policy considerations for deep learning in NLP. In: Proceedings of the 57th annual meeting of the association for computational linguistics, pp 3645–3650. https://doi.org/10.18653/v1/P19-1355

Vinuesa R, Azizpour H, Leite I, Balaam M, Dignum V, Domisch S, Fuso Nerini F (2020) The role of artificial intelligence in achieving the sustainable development goals. Nat Commun 11:233. https://doi.org/10.1038/s41467-019-14108-y

Artificial Intelligence and Big Data Analytics for Sustainable Accounting in Emerging Economies: Contemporary Approaches and Future Prospects

5

Ammar Zakaria Salem, Raed Abdelhaq, and Bahaa Awwad

Abstract

This study discusses the application of artificial intelligence in accounting analytics in developing economies, where technological change intersects with institutional and infrastructural constraints. AI technologies—automation, machine learning, and real-time auditing—are transforming accounting through improved efficiency, accuracy, and predictive ability. The study identifies key facilitators of adoption, such as digital maturity, scalable cloud-based solutions, and an innovation culture. Yet it also recognizes ongoing challenges, including regulatory fragmentation, low human capital, and nascent data governance frameworks. Country case studies on Kenya, India, and Brazil demonstrate the concrete gains of AI in improving tax compliance, fraud detection, and financial inclusion. Conversely, the study urges the importance of ethical frameworks, public–private coordination, and adaptive policies to inform implementation. Future pathways—like explainable AI, integration with blockchain, and AI-as-a-Service—are also discussed, providing practical avenues for responsible and

A. Z. Salem
Department of Administrative and Financial Sciences, Palestine Technical University—Kadoorie, Tulkarm, Palestine
e-mail: ammar.salem@ptuk.edu.ps

R. Abdelhaq
Accounting Department—Faculty of Business and Communications, An-Najah National University, Nablus, Palestine
e-mail: raed.abdelhaq@najah.edu

B. Awwad (✉)
College of Business & Finance, Ahlia University, Manama, Bahrain
e-mail: dr.awwadb@hotmail.com

B. Awwad and R. A. Awwad (eds.), *Technology and Innovation Towards a Future of Business Sustainability*, Contributions to Environmental Sciences & Innovative Business Technology, https://doi.org/10.1007/978-981-95-6469-9_5

scalable adoption. Collectively, the study provides a strategic roadmap for utilizing AI to foster transparency and resilience in financial systems throughout the developing world.

Keywords

Artificial Intelligence (AI) • Big data analytics • Sustainable development goals (SDGs) • Environmental • Social • Governance (ESG) reporting

5.1 Introduction

Artificial Intelligence (AI) has produced a revolution in the world of finance and accounting. It has affected the way the big organizations retrieve, process, and interpret the data. The inclusion of advanced AI mechanisms such as deep learning, neural networks, and robotic process automation has radically changed the approach to accounting which has been previously a rule-based method. The effect of these technologies is not only the acceleration and simplification of the processing of the data but they also facilitate the automatic detection of fraud, financial forecasting, risk forecasting, etc., so that in a word the role of the financial accountant has been expanded to more strategic (Appelbaum et al. 2017; Awwad 2024b; Hamdan et al. 2012). According to the data, AI found an increasingly sophisticated application in accounting in higher-income countries by removing the manual effort of accountants through automation, along with the growth of digital infrastructure, and the mature regulatory environment. On the contrary, in the countries with middle or low income, where AI is very less commonly used, it is expected that the advanced task of implementation will be more difficult. To explain more, those markets often face problems, such as a low level of digitalization, and the data of poor quality, leading to a lack of people who have both the knowledge to implement intelligent accounting systems as well as the competence to manage them (Alkababji and Awwad 2023; Awwad et al. 2024a; Kokina and Davenport 2017). By the way of the low-tech profit, the strong AI waves are also making their way into the local markets of Africa, Asia, and Latin America, as evidenced by the growing number of financial organizations shifting their focus to these regions as the prime candidates for a successful fintech transformation (Ozili 2023).

This study starts by exploring the function of AI in the contemporary accounting ecosystem with a focus on providing automation, fraud detection, and predictive analytics. Next the study builds upon the recognized operational constraints of technology, institutions, and regulations that shape the domain of AI in emerging economies. In developed economies, firms tend to implement AI in a structured and layered manner on existing enterprise systems. Incidentally to this gradual adoption, however, firms in emerging economies are far more likely to pursue leap-frog acceptance of AI and adopt AI solutions without employing existing legacy systems. This can exacerbate both opportunities and risks of deploying AI solutions (Salah and Awwad 2024; Vargas and Muente 2025).

The study also examined comparative AI models in accounting, including supervised and unsupervised models, and the benefits that each may bring in high and low data contexts. The study paid special attention to the hybrid intelligence of human judgement and machine computation, and how it can alleviate data scarcity and improve audit evidence, considering levels of uncertainty in regulatory circumstances (Elmegaard 2024; Razia et al. 2024). This study included several examples of real-life cases and case studies to demonstrate the application of artificial intelligence in practice, across different sectors and country contexts. The cases demonstrate how government entities, private enterprise, and the not-for-profit sector are each implementing AI technologies for tax administration, spending forecast, and automated audit work. The implementation of these projects is often in partnership with cloud AI vendors, or in an international collaboration (Hamdan et al. 2021; Rashid and Kausik 2024).

In the last section of the article, the author outlines some important upcoming trends that will influence the future of AI in accounting. These include explainable AI (XAI) as a means to provide greater algorithm transparency, embracing blockchain to improve audit tracking, and the growing reliance on AI-as-a-Service (AIaaS) to increase the number of AI tools available to accounting businesses and their clients. At the same time, the author tackles the ethical and strategic challenges of algorithmic bias, data misuse, and the potential displacement of traditional accounting roles. These challenges illustrate the need for more inclusive governance mechanisms and renewed investment in digital skills (Ikwueze 2024).

This study clearly outlines how to easily apply AI into accounting systems, from the research stage to the empirical stage, and from the policy perspective too, all while mindful of the fluid and irregular nature of new economies.

5.2 An Overview of AI in Accounting

Artificial intelligence (AI) is transforming the field of accounting by giving accountants smart tools that improve accuracy, automate tedious tasks, and support more strategic financial decisions. These tools are now a key part of any modern accounting workflow and are applied in three key areas:

- **Automation**: AI enables the automation of repeatable tasks, including data entry, processing invoices, or reconciliation processes. This increases efficiency and reduces human error.
- **Predictive Analytics**: AI systems use machine learning to evaluate historical and real-time financial information for predictions and patterns, which helps improve budgeting and longer-term financial plans.
- **Fraud Detection**: AI algorithms analyze an enormous amount of transactional data to identify anomalies and shady patterns; this provides for continuous auditing, and early detection, and prevention of fraud.

Together, these areas are changing the role of the accountant from one of transactional processing to strategic adviser, allowing organizations to utilize data in a more effective way to gain a competitive edge. The same trend supports the move toward strategic accounting and decision-making utilizing big data analytics and AI as an important feature in business (Razia et al. 2022; Thanasas and Kampiotis, 2024).

5.3 Big Data Challenges in Emerging Economies

The introduction of big data analytics in accounting in emerging economies faces a number of structural and contextual barriers. While the potential for AI-led insights is increasing, challenges continue to hinder the successful implementation of these processes:

- **Infrastructure Shortcomings**: Most emerging economies do not have sufficient digital infrastructure (reliable Wi-Fi, a modern computing environment, or cloud capabilities) to process financial data in real time.
- **Data Privacy and Security**: Weak laws and poor enforcement leave sensitive accounting data vulnerable to breaching, hacking, killing, or unethical data usage.
- **Regulatory uncertainty**: The rapid pace of technological change frequently outstrips the development of regulation, creating ambiguous or inconsistent compliance landscapes that undermines the responsible use of AI and big data tools in accounting.
- **Limited supply of skilled labor**: The limited supply of professionals who are skilled in data analytics and AI reduces the ability of organizations to adopt and use advanced accounting technologies satisfactorily.

These barriers to responsible decision-making, point to the importance of localized AI strategies that consider institutional and human-capital conditions of developing economies (Rashid et al. 2025), in order to protect against the reckless but alluring application of AI.

5.4 Current AI Applications

Despite ongoing structural challenges, several AI applications are increasingly being adopted in emerging markets to enhance accounting practices and decision-making efficiency:

- **AI Financial Forecasting**: AI tools assess historical financial information, including cash flows, trends, and market patterns, allowing for more accurate and timely financial budgets (Pawala et al. 2025).

- **Risk Profile**: Machine learning considerations can examine the financial risk and creditworthiness of clients, to assist firms with credit and investment decisions.
- **Compliance Review**: NLP systems can evaluate financial and other documents and report on violations of regulatory and compliance standards.
- **Automated Auditing**: AI capabilities provide continuous auditing and can acknowledge inconsistencies and anomalies in large data sets faster, and more accurately.

The implementation of these examples reveals an increasing attention in AI-as-a-Service (AIaaS) offerings that will help democratize intelligent tools throughout developing countries (Ikwueze 2024).

5.5 Comparative Review of AI Models

AI solutions in accounting will typically fit into one of two primary AI models, with advantages in specific situations:

- **Rule-based system**: These models function based on logical rules established by experts. They perform well in fairly stable environments and have clear regulations. They are inflexible and require manual adjustments for changes to any type of rules, laws, or standards.
- **Machine Learning (ML) Models**: These use machine learning algorithms that learn from historical data to establish patterns and make predictions. They can adapt and improve over time. They can be more precise than a static, rule-based system and are good for complex, data-intensive tasks. Their effectiveness relies on a robust and sufficient amount of quality data.

In developing economies, new hybrid models with a combination of rule-based logic, and machine learning, are attractive solutions. These models provide structure and adaptability in some resource-scarce and dynamic situations (Nasr-Azadani and Chatelain 2024).

5.6 Case Studies from Emerging Economies

- **Kenya**: A medium-sized accounting firm has utilized an AI platform to facilitate SME tax compliance that automates tax filings and data review. This has reduced the time it takes to process and review filings by 40% (Michael 2025) and improved the quality of service provided for SMEs. These use cases illustrate the encroachment of AI platforms into the financial operations of SMEs.

- **India**: A national bank adopted machine learning-based credit scoring to facilitate lending for borrowers previously excluded from the financial system. This bank used alternative data in its lending decisions to expand access to microfinance in remote rural areas, showing how AI led to a more inclusive financial system (Ikwueze 2024).
- **Brazil**: Government agencies are using AI-enabled audit tools to identify and report on financial irregularities in real time. The use of these tools has cut losses from fraud-related errors for public sector reporting by more than 25% and supports initiatives aimed at strengthening transparency in public financial accountability. The successful implementation of these technologies illustrates the potential of AI for enhancing accountability (Rashid et al. 2025).

5.7 Future Trends in Accounting and AI

There are several transformative trends that are likely to reshape the future of accounting work with AI—particularly in terms of the industries that work in developing and resource-constrained economies.

- **Blockchain Integration**: The integration of blockchain with AI is enabling more secure and transparent accounting practices. It has been demonstrated that blockchain's ledgers are immutable, while intelligent systems can validate transactions for audibility with traceability. In this instance, accounting work holds significant promise for reducing fraud and improving accountability within both private and public sector financial management (Nasr-Azadani and Chatelain 2024; Razia and Awwad 2022).
- **Explainable AI (XAI)**: As AI systems become more complex, the demand for interpretability has grown. We need to be able to rely on transparent and human-understandable reasons for the algorithmic decisions that are produced by explainable AI. This is an important requirement for financial reporting and audit for accountability and regulatory purposes (Pawala et al. 2025).
- **Real-Time Auditing**: AI applied technology has ushered in the capability for continuous auditing through live streams of financial data. This ensures real-time anomaly detection and supports proactive risk management, which is critical in the fast-paced or high-volume environments often found in large organizations and government agencies (Michael 2025).
- **AI-as-a-Service**: The emergence of new cloud-based AI platforms is increasing exponentially to extend small and medium-sized enterprises (SME's) access to scalable and affordable solutions. These AI strategies provide firms without substantial technology infrastructure the ability to engage advanced tools to provide financial forecasting, fraud detection, and compliance monitoring. (Ikwueze 2024) (Fig. 5.1).

Together, these innovations point to a future where AI-driven accounting achieves greater accuracy and efficiency, combined with greater access, accessibility, and

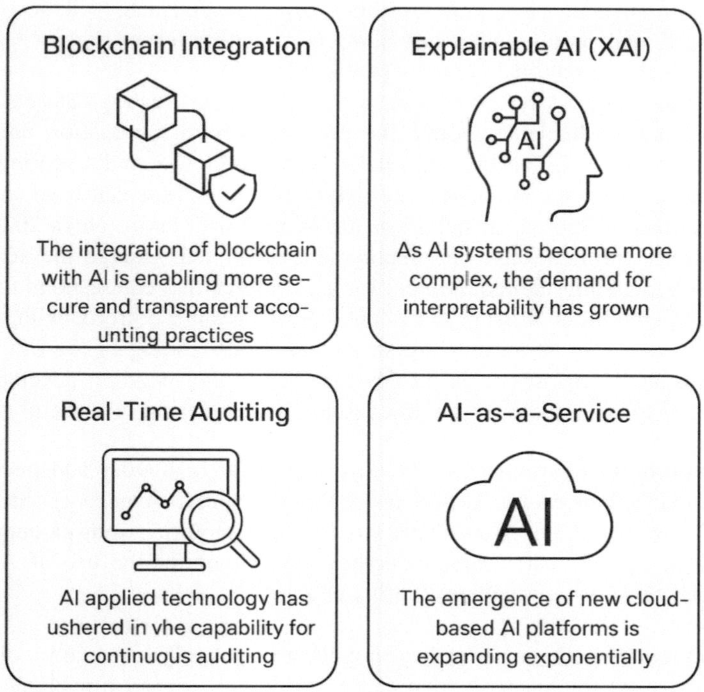

Fig. 5.1 Future trends in accounting and artificial intelligence (AI)

flexibility for organizations in developing economies. The convergence of these new technologies suggests the potential to create more resilient, transparent, and inclusive financial systems applied across emerging economies.

5.8 Policy and Ethical Considerations

As artificial intelligence (AI) continues to permeate accounting systems in, especially in developing economies, it will be important to explore a few different policy and ethical issues that can govern and facilitate responsible and sustainable use of AI in accounting contexts because the absence of these policies and frameworks could lead to unintentional damage, systemic bias, or the loss of public trust.

- **Data Governance**: Ethical AI requires proper data governance. Organizations must communicate clear policies about how financial data will be collected, processed, and stored, and ensure compliance with relevant privacy legislation in local and international contexts (Awwad 2024a).
- **Challenge Bias**: AI systems must be trained on constructive datasets that are diverse and equitable to reduce the potential for discrimination and biased decision-making. In accounting-related fields such as credit scoring, auditing, and regulatory compliance, the risk of biased outputs can have significant implications, including financial loss and reputational harm (Nuka 2024).
- **Workforce Implications**: The deployment of AI will change the accounting function, including the requirement for a workforce that can respond to the use of new digital tools. It is not just about equipping our workforce with technical training; it is also about creating an ethical understanding of the limits of AI and how they can impact the profession. Supporting ongoing development is important to allow accountants to continue to be relevant and useful (Ikwueze 2024).
- **Accountability Frameworks**: We will need clear regulatory and institutional approval mechanisms for AI decision-making such that it can be traced, audited, and challenged if necessary. This would mean requiring transparency of the algorithm designs and oversight bodies which monitor the use of AI in the financial services environment (Farina et al. 2025; Nuka 2024).

Incorporating these ethical and policy safeguards is not only a regulatory obligation but also essential for fostering trust and ensuring the long-term resilience of AI-driven accounting systems.

5.9 The Role of Digital Maturity in Shaping AI Adoption in Accounting

Organizations with more established digital landscapes (e.g., ERP systems, cloud-based accounting systems, etc.) are better suited for the adoption and scaling of AI technologies (Prasad 2024; Taribagil et al., 2023).

- **System Compatibility:** For organizations with a cohesive digital ecosystem, integrating AI tools into their operations is less technical. When existing systems are compatible (environment, systems), it is easier to automate workflows and sync data (Prasad 2024).
- **Organizational Culture:** Businesses that have reached digital maturity often have cultures that encourage a focus on innovation, experimentation, and collaboration among cross-functionality. This attitude toward forward-thinking culture tends to facilitate acceptance and adoption of AI technologies in-house (Taribagil et al. 2023).

- **Data Readiness:** Digitally developed organizations typically have strong data governance and data quality systems. In other words, organizations with more digital maturity usually have clean, well-structured, and accessible data. The importance of good data is key for developing AI models that produce reliable outputs and accurate analysis outcomes (Alex-Omiogbemi et al. 2024).

For this reason, building digital maturity is a strategic pre-requisite for AI in accounting, especially in emerging markets where each organization may have widely varying degrees of (digital) readiness.

5.10 Cost–Benefit Analysis of AI Implementation in SME Accounting

AI adoption comes with significant financial and operational implications for SMEs. While the costs of implementation might be high, increasing access to AI-as-a-Service (AIaaS), and expanding digital platforms with scalable solutions, makes intelligent solutions more readily accessible and more easily implemented (Prasad 2024; Alex-Omiogbemi et al. 2024; Taribagil et al. 2023).

- **Upfront Cost vs Long-Term Benefit:** The costs of AI implementation might be a hurdle at first, but it can also bring down future operational costs in the long run when implemented. Automating accounting tasks that are performed manually, such as data entry and reconciliation, leads to improved efficiency, a reduction in mistakes, and a potential for longer-term cost control.
- **Access to Scalable Tools:** Cloud-based tools allow SMEs the opportunity to implement technology based on their scale, which reduces the need for costly in-house technology infrastructure and allows them to grow their digital capacity incrementally.
- **Competitive Advantage:** Organizations that are early adopters of ways to use AI get a quicker monthly financial report cycle, a better compliance accuracy, and better decision-making information for support. These functions improve internal processes and create increased credibility and trust with clients and regulators.

In summary, AI adoption can create sustainable value and operational resilience with strategic clarity in an increasingly competitive environment.

5.11 Institutional Barriers to AI Integration in Developing Countries

While the technical potential of AI in accounting is widely recognized, developing countries often face entrenched institutional barriers that slow or hinder its adoption. Institutional barriers go beyond infrastructure and finance and include fragmented laws and policies, institutional inertia, and policy inconsistencies. As traditional regulatory systems are disrupted by AI and large language models, the lack of institutional readiness in developing countries becomes an additional barrier to AI adoption. (Kothandapani 2025).

- **Regulatory Inconsistency:** Many developing economies find themselves unable to update financial and technology regulations or provide consistency in the application of those laws. All firms need is uncertainty for future investment in AI innovation, and creating intelligent systems.
- **Capacity Challenges:** Public institutions often do not have the technical skills in-house to ensure the supervision and oversight of AI use. Without personnel with the skills to even check algorithmic systems, governments have little hope of determining a safe, ethical, and effective process of AI adoption.
- **Innovation Resistance:** Structures and historic hierarchies, and risk aversion in the institutional culture of public finance and historic accounting organizations inhibits the change associated with disruptive technologies at a slower pace and prevents appropriate change in time.

Addressing the institutional barriers to developing AI is about creating a proactive policy environment for the use of responsible AI everywhere: modernizing regulation and compliance, institutional capacity building for supervision and oversight in public finance institutions and policymaking that promotes or enables innovation-friendly governance models and development of responsible and relevant AI applications across financial systems.

5.12 Benchmarking AI Readiness: A Comparative Regional Analysis

Understanding a country's level of AI readiness is critical for formulating policy directions, allocating resources, and guiding digital transformation strategies, particularly in accounting. Regional comparisons reveal significant variations in the quality of infrastructure, institutional capacity, and workforce readiness—all of which impact the speed and effectiveness of AI adoption in financial systems.

- **Infrastructural and Connection Quality:** Regions with advanced telecommunications networks and established and good cloud infrastructure, such as Southeast Asia, have adopted AI applications more quickly. Those regions facing low-quality connectivity, including some in sub-Saharan Africa, can anticipate delays in adopting AI.
- **Policy and Governance:** Where governments have developed clear national AI strategies, provided flexible regulatory frameworks (e.g., UAE, India), countries demonstrate higher rates of adoption. Governments support collective certainty through stability in governments intent to disrupt sectors and segment industries.

- **Human Capital Development:** Education systems that include digital literacy, data science, and accounting analytics provide strong bases for AI readiness and promote the role of public universities and vocational institutions preparing the next generation of accountants with AI talent (O'Neill and Bagchi-Sen 2023).

These dimensions determine the speed with which industries can benefit from the transformation of AI in accounting. Clearly, a multi-stakeholder approach, taking into account infrastructure development, policy reform, and human-capital development, is required to advance this agenda.

5.13 Strategic Recommendations

To enable emerging economies to leverage the full potential of AI for accounting, there should be a comprehensive strategy addressing both institutional preparedness and technological competence. The following suggestions offer top priority areas of integrating AI that are sustainable and equitable:

- **Invest in Infrastructure**: Enhancing digital infrastructure—primarily broadband access, cloud storage, and secure data centers—is the backbone of deployable AI at scale. Tier III and Tier IV data center investments are especially most vital to facilitate stable financial data operations (Abiodun 2025).
- **Strengthen Regulatory Frameworks:** Governments need to develop agile, future-oriented legal frameworks that adapt to the advancements in technology. Innovative and clear regulation not only promotes innovation but also ensures data integrity and establishes trust in AI systems (Islam et al. 2023).
- **Establish Public–Private Partnerships:** Intersectoral coordination among state governments, technology providers, and educational institutions is essential in tackling local AI capacity. Cross-sectoral partnerships assist in closing the skills gaps, localizing solutions, and speeding up accountable innovations (Islam et al. 2023).
- **Reform Education and Training**: Infusing AI, data analytics, and ethics into accounting and business education will provide future professionals with key digital competencies. Continuing professional development must also target current practitioners to aim for widespread-level adoption.
- **Enable Experimental Environments:** The creation of AI sandboxes, living labs, and testbeds provides a space to test AI tools in a controlled environment. Such environments facilitate hands-on experimentation, evidence-based policy decisions, and secure testing of finance-centered AI applications (Schuurman et al. 2024).

Collectively, these strategic pillars outline a plan for the implementation of AI in accounting by emerging markets in a manner that is contextually relevant, sustainable, and consistent with long-term development aspirations.

5.14 Conclusion

The integration of artificial intelligence into accounting practices in emerging economies demonstrates tremendous potential, as well as ongoing structural challenges. While AI technologies, including automation, predictive analytics, and continuous auditing, will enhance financial flows, it is important to note that the success of these products depends on organizations' technological, organizational, and cultural readiness to implement them. (Prasad 2024; Taribagil et al. 2023). The dimension of digital maturity has been highlighted already in this study, but it is a major component nonetheless. Organizations that have established digital capabilities and data governance are likely better positioned to adopt and scale AI solutions effectively (Alex-Omiogbemi et al. 2024). In contrast, small- and medium-sized enterprises (SMEs) need to consider the cost–benefit relationship in terms of adoption costs when factoring in long-term efficiency, regulatory compliance, and competitiveness (Prasad 2024; Taribagil et al. 2023).

At the policy level, barriers at the institutional level (e.g., fragmented regulations, weak enforcement, and lack of public-sector technical capacity) still obstruct AI integration (Kothandapani 2025). On top of this, and from the perspectives of comparative analysis, clear variations in infrastructure, and human-capital development, as well as national strategies on AI, imply that, regionally, there would need to be interventions in education, telecommunications, and regulatory experimentation (Uzoka et al. 2024; O'Neill & Bagchi-Sen 2023; Schuurman et al. 2024).

Case studies discussed in this study reinforce that AI can enhance transparency, tax compliance, and fraud detection if deployed under appropriate institutional conditions (Michael 2025; Ikwueze 2024; Rashid et al. 2025). Yet, such outcomes will only materialize dependent on responsible governance, ethical safeguards, and continued investment in digital skills and public–private collaboration.

In summary, realizing the potential benefits of AI in accounting across emerging economies will require a more integrated approach than simply adopting new technology; it will require efforts to coordinate infrastructure, change regulation, and to further people development. If nations integrate digital policies prepared with care, AI can become a constructive agent of change to help build more resilient, more efficient, and equally developed financial systems.

References

Abiodun K (2025) Digital infrastructure and sustainable data centers investment in Africa: role of Tier III & Tier IV

Alex-Omiogbemi AA, Sule AK, Omowole BM, Owoade SJ (2024) Conceptual framework for advancing regulatory compliance and risk management in emerging markets through digital innovation

Alkababji MW, Awwad BS (2023) Big data tools and techniques in developing accounting practices and the impact of their analysis on the financial performance. Central Eur Manage J 31(1):618–631

Appelbaum D, Kogan A, Vasarhelyi M, Yan Z (2017) Impact of business analytics and enterprise systems on managerial accounting. Int J Account Inf Syst 25:29–44
Awwad BSAL (2024a) Governance with relationship between entrepreneurship and economic growth in Palestine. Int J Law Manage 66(2):259–287
Awwad B (Ed) (2024b) The AI revolution: driving business innovation and research: volume 1, vol 524. Springer Nature
Awwad BS, Alkababji MW, Razia BS (2024) Artificial intelligence and the quality of accounting information in Palestinian industrial companies. J Financ Rep Account (ahead-of-print).
Elmegaard J (2024) Navigating the digital frontier in accounting: transformative impacts and interplay of digitalization, accounting systems, and management accountants in an institutional context. Copenhagen Business School [Ph.D.]
Farina M, Yu X, Lavazza A (2025) Ethical considerations and policy interventions concerning the impact of generative AI tools in the economy and in society. AI Ethics 5(1):737–745
Hamdan A, Hassanien AE, Khamis R, Alareeni B, Razzaque A, Awwad B (eds) (2021) Applications of artificial intelligence in business, education and healthcare, vol 954. Springer, Cham
Hamdan AMM, Kukrija G, Awwad BSA, Dergham MM (2012) The auditing quality and accounting conservatism. Int Manage Rev 8(2).
Ikwueze C (2024) Artificial intelligence as a service: a systematic review of literature from the customer's perspective
Islam R, Patamsetti V, Gadhi A, Gondu RM, Bandaru CM, Kesani SC, Abiona O (2023) The future of cloud computing: benefits and challenges. Int J Commun Netw Syst Sci 16(4):53–65
Kokina J, Davenport TH (2017) The emergence of artificial intelligence: how automation is changing auditing. J Emerg Technol Account 14(1):115–122
Kothandapani HP (2025) AI-driven regulatory compliance: transforming financial oversight through large language models and automation. Emerg Sci Res 12–24
Michael O (2025) Maximising the potentials of small and medium scale business enterprises in developing nations through the use of artificial intelligence: AI adoption by SMEs in the developing nations. In: The future of small business in industry 5.0. IGI Global Scientific Publishing, pp 215–246
Nasr-Azadani MM, Chatelain JL (2024) The journey to trustworthy AI-part 1: pursuit of pragmatic frameworks. arXiv preprint arXiv:2403.15457
Nuka ST (2024) Exploring AI and generative AI in healthcare reimbursement policies: challenges, ethical considerations, and future innovations. Int J Med Toxicol Legal Med 27(5):574–584
O'Neill M, Bagchi-Sen S (2023) Public universities and human capital development in the United States. GeoJournal 88(1):733–751
Ozili PK (2023) Impact of monetary policy on financial inclusion in emerging markets. J Risk Fin Manage 16(7):303
Pawala T, Phunsa S, Kenaphoom S (2025) Utilizing AI for cash flow forecasting: enhancing predictive accuracy in business finance. In: Sustainable futures with life cycle assessment in industry 5.0. IGI Global Scientific Publishing, pp 423–440
Prasad S (2024) Artificial intelligence and automation in accounting: opportunities and challenges for the future. Unified Visions, p 306
Rashid A, Baloch N, Rasheed R, Ngah AH (2025) Big data analytics-artificial intelligence and sustainable performance through green supply chain practices in manufacturing firms of a developing country. J Sci Technol Policy Manage 16(1):42–67
Rashid AB, Kausik AK (2024) AI revolutionizing industries worldwide: a comprehensive overview of its diverse applications. Hybrid advances, p 100277
Razia B, Awwad B, Taqi N (2022) The relationship between artificial intelligence (AI) and its aspects in higher education. Dev Learn Organ: Int J 37(3):21–23
Razia AS, Alkababji MW, Ruzieh AS, Awwad BS, Razia BS (2024) Competitive advantage in light of AI to achieve innovative as a key of the SDGS: Palestinian insurance industry. J Lifestyle SDGs Rev 4(1):e01870–e01870

Razia B, Awwad B (2022) A comprehensive review of blockchain technology and its related aspects in higher education. Technologies, artificial intelligence and the future of learning post-COVID-19: the crucial role of international accreditation, pp 553–571.

Salah AS, Awwad BS (2024) Intelligence and tax compliance. Achieving sustainable business through AI, technology education and computer science: volume 3: business sustainability and artificial intelligence applications, vol 163, p 53

Schuurman D, De Cock L, Canter M, De Meester T, Willems O, De Swert T (2024) Testing and experimentation facilities: exploring the link with AI regulatory sandboxes, living labs and AI testbeds. In: OpenLivingLab days conference 2024, p 222

Taribagil P, Hogg HJ, Balaskas K, Keane PA (2023) Integrating artificial intelligence into an ophthalmologist's workflow: obstacles and opportunities. Expert Rev Ophthalmol 18(1):45–56

Thanasas GL, Kampiotis G (2024) The role of big data analytics in financial decision-making and strategic accounting. Technium Bus Manage 10:17–33

Uzoka A, Cadet E, Ojukwu PU (2024) The role of telecommunications in enabling internet of things (IoT) connectivity and applications. Compr Res Rev Sci Technol 2(02):055–073

Vargas F, Muente A (2025) Artificial intelligence framework for the inter-American development group

Part II
Green Economy and Sustainable Policies

Resilient Cities Under Occupation: Localizing Doughnut Economics in Ramallah, Palestine

6

Alicia S E Bahbah

Abstract

This chapter studies the potential application of Doughnut Economics in the Global South through a case study of Ramallah in the occupied Palestinian territory, which is known for its rapid urbanization and political instability. Although the Doughnut Model provides a framework for creating a "safe and just space for humanity," its implementation in conflict-affected areas remains overlooked. This chapter critically examines how Ramallah has started localizing some aspects of this model by building on the downscaling approach, which highlights the role of local governance and community-based projects in applying such a model. It explores how Ramallah integrated smart city initiatives and sharing platforms projects to adapt global ecological economics models to its own political and socio-economic context. Therefore, arguing that despite the systemic limitations, Ramallah's hybrid strategy offers a grounded example of how cities under occupation can still develop an innovative framework to promote urban sustainability and social equity.

Keywords

Doughnut economics • Urban sustainability • Local governance • Smart cities • Anthropocene

A. S. E. Bahbah (✉)
Corporate, Commercial and Taxation Law, University of New South Wales, Sydney, Australia
e-mail: abahbah@ittqanlaw.com

B. Awwad and R. A. Awwad (eds.), *Technology and Innovation Towards a Future of Business Sustainability*, Contributions to Environmental Sciences & Innovative Business Technology, https://doi.org/10.1007/978-981-95-6469-9_6

6.1 Introduction

Since the Industrial Revolution, humanity has entered the *Anthropocene.* This epoch is characterized by significantly exacerbated both environmental degradation and social inequality, which have exposed the vulnerability of the global system (Hunter 2019). Five of the nine identified planetary boundaries have already been exceeded (Khoday 2024), particularly across the Global South (Rockström et al. 2009). These socio-ecological issues are often traced back to the dominating global economic paradigm, driven by neoclassical economic theories. Grounded in principles of perpetual growth and the relentless pursuit of capital accumulation, often referred to as "growthism" (Schneider et al. 2010). Beyond its economic dimensions, this system has played a critical role in contributing, in many cases, to geopolitical instability and colonization, including the Israeli occupation of Palestinian territory.

These tipping points prompted a critical examination of the current capitalist mechanisms. Transformative economics "post-growth" or "ecological economics" models emerged as a response (McAlpine et al. 2015). One notable model is the Doughnut Economics Model, which proposes a safe and just space for humanity, balancing human needs with planetary boundaries through sustainability-focused shifts, emphasizing redistributive and regenerative practices (Raworth 2018). This model has been adopted in cities such as Amsterdam.

However, Doughnut Economics focuses on advanced economies (Vandenhole 2018). There remains limited exploration on implementing this model in conflict-affected, marginalized, and developing urban areas (Karimabadi 2024). Palestine presents a unique case in this matter, as the ongoing Israeli occupation restricts resource sovereignty and human rights violations, compounded by internal governance fragmentation, collectively posing serious barriers for implementing such models (United Nations Country Team in the Occupied Palestinian Territory 2022). Yet, this case also opens up new questions: Can models like Doughnut Economics be meaningfully localized or adopted under such conditions? And if so, what Kinds of legal instruments and governance innovations are needed to enable such localisation?

Given that cities play a central role in addressing the challenges of the Anthropocene (Aust and Nijman 2021). This chapter addresses these questions by focusing on the city of Ramallah, the de facto administrative center of the West Bank, a city of both resilience and contradiction. It also explores how the principles of Doughnut Economics could be adapted through legal innovation, commons-based governance, and smart city initiatives for a transformative, sustainable city under constraint.

6.2 Methodology

This study adopts a qualitative analysis approach by examining and exploring key literature related to Kate Raworth and other ecological economics pioneers such as Jason Hickel and David Bollier. Data and information were gathered from secondary sources, including institutional reports, published strategies from relevant international and national organizations, and peer-reviewed journal articles from major databases like JSTOR and ScienceDirect. Moreover, the study included direct interaction with Ramallah Municipality employees, during which knowledge was shared regarding the ongoing projects and initiatives. This study also focuses on comparative analysis by examining how Doughnut Economics has been adapted in other cities, and contrasts these applications with the unique challenges posed by Israeli occupation in Palestine. Overall, this analysis is theoretical and interpretive, aiming at understanding how global sustainability models can be applied through locally grounded approaches.

6.3 Literature Review

6.3.1 The Doughnut Model: Bridging Theory with Reality in Palestine

The Doughnut Economics is represented as a doughnut-shaped diagram, with two distinct boundaries. The inner boundary, "the social foundation" symbolized by the central hole, highlights areas of deprivation where individuals lack necessities due to traditional economic system flaws. In contrast, the outer boundary, "the ecological ceiling," marks the planetary limits that must not be crossed to prevent environmental degradation (Hickel 2019, p. 19). Together, these boundaries define a "safe and just space for humanity," integrating social and ecological dimensions into a holistic approach to sustainability (Ahlström 2019) (Fig. 6.1).

Regarding its principles and practices, the Doughnut Economics Model emphasizes seven shifts that include distributive and regenerative practices (Doughnut Economics Action Lab 2024).

This paradigm acknowledges that sustainability cannot be merely an environmental concern; it rather asserts that sustainability requires the integration of social needs, welfare, and economic opportunities within the ecological limits, therefore creating a truly sustainable society (Agyeman 2013, p. 5). Consequently, the social foundation ensures that no one falls short of life's essentials, including access to sufficient food, clean water, sanitation, and more. This must be achieved through equality, justice, and urban sustainability commitments. However, in many countries, the social foundation remains unmet (Fanning et al. 2022), and there is a huge lack of equality and justice, which is the foundation of this model, with Palestine serving as a stark example.

In Palestine, Raworth's vision of social foundation is profoundly undermined by the Israeli occupation. According to the 2022 Common Country Analysis (CCA),

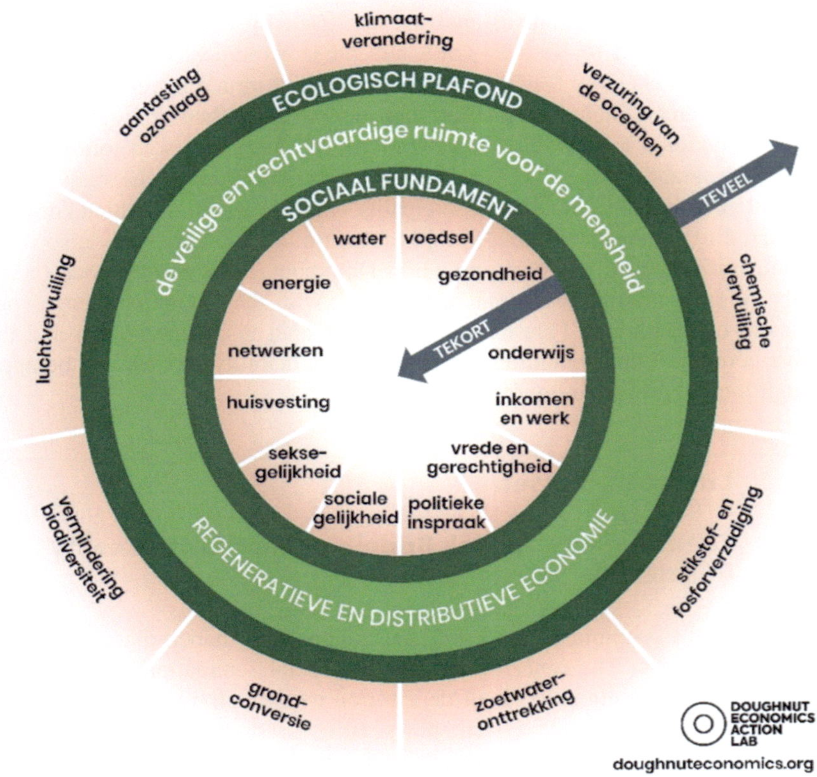

Fig. 6.1 Doughnut economics model: balancing environmental limits and social foundations *Source* Doughnut Economics Action Lab (2024)

the occupation in Palestine, coupled with internal political divisions, serves as the primary barrier to achieving sustainability at all levels (United Nations Country Team 2022, p. 24). It restricts access to essential resources, including land, water, and energy, and systematically denies Palestinians their fundamental needs, thus undermining social welfare and challenging the efforts to develop cohesive and resilient urban systems (Khoffash 2024, p.2).

This reality highlights the significant gap between the ideals of Doughnut Economics' social foundation and the harsh facts in Palestine. The Palestinian Authority has acknowledged that achieving long-term economic and social sustainability is extremely challenging, if not impossible, under the current conditions (State of Palestine 2016, p. 27). This literature analysis reveals the pressing need for structural changes to address the status quo. Achieving urban sustainability and social justice through Doughnut Economics requires innovative approaches, including prioritizing collaboration with local governments, enhancing the efficiency of urban systems, and adopting sharing economy models rather than relying solely on the Palestinian Authority as the primary governing body.

6.4 The Role of Local Governance

Local governance has long been recognized as pivotal for achieving sustainability and in addressing planetary boundary concerns posed by the Anthropogenic. This is due to their proximity to communities and their capacity to address context-specific challenges (Aust and Nijman 2021). Raworth underscores the critical role of local governance through local governments in implementing the doughnut economics framework. Local governments are more likely to address complex, interconnected challenges in a manner that responds directly to local needs (Raworth 2018, p. 59). Whereas, the importance of local governance stems from the absence of cohesive global or national framework guidelines for implementation (Turner and Wills 2022, p. 8). Moreover, due to the limitation of the interstate system and state sovereignty in responding effectively to planetary boundaries (Aust and Nijman 2021, p. 105).

Consequently, policymakers continually argue that local governments must take the initiative to build momentum through resilient approaches that inclusively engage a wide range of stakeholders. To be effective, local initiatives should integrate their frameworks within the existing policies and governance structure to serve as an impactful tool for addressing socio-ecological issues (Turner and Wills 2022, p. 8).

This approach is called "Downscaling Approach," where local governments are adapting global models and frameworks to smaller, more localized levels such as cities or communities (Aust and Nijman 2021, p. 103). In light of the exceptional situation, alongside the absence of central sovereignty of the Palestinian government. This legal approach could be seen as the most applicable one to deal with the status quo and take the initiative to implement doughnut economics in Palestine.

Following the establishment of the Palestinian Authority in 1994 under the Oslo Peace Agreement, planning responsibilities were transferred to Palestinian institutions, with the Ministry of Local Government (MOLG) and other bodies tasked with key administrative functions. In the absence of a stable and strong central government, Local Government Units (LGUs) have become the backbone of public administration, shouldering immense responsibilities to meet individuals' basic needs such as water and electricity supply, and urban planning, therefore, promoting community welfare (Shaikh Ali and Khalid n.d., p. 3). This link between local governance and economic dynamism is also evident in literature discussing how entrepreneurship contributes to economic resilience in contexts like Palestine (Awwad 2024). Consequently, local governance remains the cornerstone for state-building and democratization, offering a critical pathway for addressing context-specific needs and fostering resilience in Palestinian communities (Abdelhamid 2006, p. 8). However, this chapter focuses on examining the Ramallah Governorate.

6.5 "Resilient Ramallah 2050"

Ramallah governorate stands as the most rapidly growing urban center in the West Bank, serving as the administrative hub of the Palestinian Authority and a capital for public and private sector services, including health, financial, corporations, and educational institutions (AbuSada and Thawaba 2011, p. 383). His centralisation has transformed Ramallah into a pivotal city for urban sustainability studies and an ideal setting to examine the practical application of Doughnut Economics. Consequently, this swift urbanization has attracted significant foreign investment and non-profit initiatives, particularly in green projects and endeavors aimed at social and economic welfare (World Bank Group 2021). For instance, the Italian Agency for Development Cooperation, AICS, is currently funding the local authorities, mainly the Ramallah governorate, to support them in adopting a comprehensive sustainable development framework and practices. This project aims to transfer the Italian experience to local government to strengthen environmental frameworks, promote circular economy principles, and foster the regeneration of green spaces within urban areas (Association of Palestinian Local Authorities [APLA] 2024; Khoffash 2024).

Another notable example by which Ramallah City is demonstrating its efficiency in implementing more urban planning projects, thus leading the city to a sort of "*sweet spot*," is The *Resilient Ramallah 2050* strategy. This strategy was developed by Ramallah municipality as part of the *100 Resilient Cities* initiative, which seeks to integrate global best practices with local governance to address Ramallah's unique challenges (Ramallah Municipality 2017, p. 18). Through this project, Ramallah has become linked with other major cities worldwide, therefore facilitating resource exchange, information sharing, and cooperative problem-solving. For example, cities like Rotterdam in developing infrastructure partnerships and Byblos in conserving cultural heritage have served as models for Ramallah (Ramallah Municipality 2017, p. 15).

Building on the downscaling approach, technological advancements and innovative urban models have opened new avenues for addressing the city's challenges. Among these, the concept of smart cities and the integration of sharing platforms present promising solutions for fostering urban sustainability and enhancing community resilience (Trindade et al. 2017, p. 2). These approaches can serve as a crucial tool for Ramallah in overcoming its socio-economic constraints while promoting a more inclusive and sustainable urban environment.

6.6 "Ramallah Smart City: Freedom Through Technology"

Over the past two decades, Ramallah City has experienced rapid growth in population and economic development, leading to an increasing demand for improved services to meet the needs of its residents (AbuSada and Thawaba 2011, p. 382). However, due to the political and security constraints, the pressing need to utilize technology to empower the community and enhance the quality of life arose. In this response, Ramallah Municipality launched the Ramallah Smart City (RSC) project, in collaboration with various stakeholders, including the private sector and social technology enterprises, supported by international green funds (Urban Agenda Platform n.d.). This transformative project sought to enable and enhance social qualities for both residents and visitors, achieving multifaceted pillars such as smart infrastructure, smart governance, smart education, and smart environment (Jaber and Burbar 2023, p. 476). Such initiatives, when coupled with transparent environmental disclosure, enhance community trust and reduce information asymmetries (Awwad 2022).

The project came under the title of "Freedom Through Technology," which has broken many restrictions imposed on Ramallah residents. Despite limitations such as the Israeli ban on importing 4G technology and other critical equipment, Ramallah Municipality implemented creative solutions to transform the city. For example, all major streets in Ramallah were turned into Wireless Access Zones, offering free Wi-Fi and access to municipal e-services all the time. Moreover, technology-based educational tools were introduced to public schools, with one of these becoming the first "Smart School" in Palestine (Jaber and Burbar 2023, p. 476). These digital transformation efforts have indeed contributed to long-term population well-being, making Ramallah city the first smart city. Moreover, it enabled citizens to access municipal services remotely and facilitate education for students despite movement restrictions and beyond.

The concept of smart cities is relatively modern, emerging alongside advancements in technological innovations and evolving urban development models, including information cities, digital cities, and sustainable cities (Yigitcanlar 2006). Nevertheless, smart cities heavily rely on information and communication technology (ICT) to enhance their competitive advantages or can be viewed as a framework for urban development that integrates human, social, and technological resources (Angelidou 2014), Such concerns highlight the urgent need for integrated frameworks that merge governance with artificial intelligence to ensure equitable digital development (Awwad 2025).

However, these initiatives are becoming increasingly linked to platform capitalism, particularly when tech initiatives are dominated by the private sector, by which they transform data into private assets and monopolize digital infrastructure (Srnicek and De Sutter 2017, p. 23). Given the sensitive political landscape in Palestine, the utilization of tech platforms requires a high level of data governance oversight (Awwad and El Khoury 2024). Data governance must be carefully addressed to prevent data exploitation by private tech enterprises (Artyushina

2020). This issue is crucial in the Palestinian context, where the misuse of data could exacerbate existing vulnerabilities and further entrench systemic injustices.

In response, the second direction of the Ramallah Municipality Strategy, "Responsive Governance," came to emphasize the enhancement of governance frameworks. One significant measure involves centralizing data to a certain degree while ensuring full transparency and accountability to the public (Ramallah Municipality 2017, p. 46). Moreover, it is essential to implement legal instruments that ensure data generated within the locality is owned and controlled by the community (Johnson et al. 2022, p. 4). To tackle the potential threats posed by platform capitalism and exploit data, and to support the Palestinian initiatives away from capital flow, these issues could be addressed through legal innovations and community-based frameworks as they become crucial. Such approaches can ensure that the Smart City project in Ramallah remains a tool for social empowerment and community welfare, rather than a mechanism for exploitation.

6.7 Commons and Legal Hacks for Fair Smart City

The rise of platform capitalism is deeply rooted in the current legal systems, which are often shaped to support a capitalist agenda that prioritizes profit maximization over social welfare. This dynamic is exacerbated by weaknesses of consent-based data governance and the selective enforcement of privacy regulations. To tackle these challenges, community-driven approaches like commoning and innovative strategies that promote shared commons in society should be adopted, therefore reclaiming more resources and promoting equity (Bollier 2007, p. 12).

To support commoning practices under the current legal system, Bollier refers to "legal hacks" as creative strategies that use existing laws in innovative ways to provide solutions that serve the public good and support community initiatives (Bollier 2020, p. 225). For instance, copyright laws could be employed to create more participatory and free platforms that contribute to improving the quality of life for society as a whole, rather than relying on platform capitalism. A compelling example is the General Public License (GPL) developed by Richard Stallman, which allows software to be free and available to use, modify, and redistribute without any restrictions (Bollier 2020, p. 234). Such collaborative models could be scaled through the active participation of national investment entities like the Palestinian Investment Fund, which has shown alignment with social responsibility principles (Awwad et al. 2023).

In Ramallah, such legal innovation could be used in the digital infrastructure in the Smart City project, to establish data trusts managed collaboratively by citizens, empowering communities to decide how data is used to support public services and social welfare. Likewise, public digital resources may be managed using commons-based frameworks that promote inclusivity and collective accountability. Moreover, this approach could be utilized to empower the educational system, as the Smart City project aims to develop open-source educational platforms, owned and operated by Palestinians. Such platforms would enable knowledge sharing, promote community interaction, and mitigate the exploitative impact of the Israeli occupation (Jaber and Burbar 2023, p. 476). These efforts align with Kate Raworth's concept of distributive justice, assuring a more equitable distribution of resources, knowledge, and opportunities.

In this regard, the interpretation of social enterprises is crucial in reinforcing a commons-based approach. Palestinian social enterprises play a pivotal role in supporting initiatives launched by the Ramallah Municipality. For example, these enterprises can participate in developing open-source educational platforms (Amr 2023). On the environmental side, it is noted that non-profit green projects that come under the social enterprises have witnessed a significant rise in Palestine since the launch of the "Ramallah Smart City" project (SwitchMed 2022). These projects contribute to providing society with less expansive renewable energy projects such as solar panels, therefore reducing dependence on Israeli electricity companies, and extending electricity to marginalized areas.

Integrating these efforts with the Doughnut Economic framework can contribute to creating a "safe and just" environment for Palestinian society. Ultimately, advancing distributive justice by ensuring equitable access to education and services while fostering environmental sustainability through green initiatives.

6.8 Conclusion

This case study emphasizes that the Doughnut Economics model can indeed be localized in occupied and marginalized contexts. Ramallah's experience with smart city initiatives, sharing platforms, and legal innovation demonstrates how locally based strategies can still provide a practical foundation for implementing Doughnut Economics, even under the constraints posed by Israel. The city's efforts demonstrate its capacity to reimagine sustainability in the Anthropocene, establishing itself as a symbol of resilience and justice among systemic challenges. However, the effective implementation process depends on a holistic approach that integrates strong local governance and community-led initiatives. Through the promotion of a culture of shared responsibility and sustainability, Ramallah can evolve into a model city that embodies the vision of a "safe and just space for humanity." These efforts provide a valuable precedent for other cities in the Global South exploring pathways to sustainability.

References

Abdelhamid A (2006) Urban development and planning in the occupied Palestinian territories: impacts on urban form. In: Conference on Nordic and international urban morphology: distinctive and common themes, pp 1–17

AbuSada J, Thawaba S (2011) Multi criteria analysis for locating sustainable suburban centers: a case study from Ramallah Governorate, Palestine. Cities 28(5):381–393

Agyeman J (2013) Introducing just sustainabilities: policy, planning, and practice. Zed Books Ltd

Ahlström H (2019) Policy hotspots for sustainability: changes in the EU regulation of sustainable business and finance. Sustainability 11(2):499

Amr L (2023) Can social entrepreneurship save Palestine? Harnessing social entrepreneurship for economic and social development in Palestine. Week Palestine 301:14

Angelidou M (2014) Smart city policies: a spatial approach. Cities 41:S3–S11

Artyushina A (2020) Is civic data governance the key to democratic smart cities? The role of the urban data trust in Sidewalk Toronto. Telematics Inform 55:101456

Association of Palestinian Local Authorities (2024) L.A.N.D project: local authorities network for sustainable development [Web page]. https://www.apla.ps/en/projects/4551.html

Aust HP, Nijman JE (2021) Planetary boundaries intra muros: cities and the Anthropocene. In: Research handbook on law, governance and planetary boundaries. Edward Elgar Publishing, pp 103–123

Awwad BS (2022) The role of environmental disclosure in enhancing the market value of industrial and service companies listed on the Palestinian stock exchange. Int J Entrepreneurship 26(2):1–12. ISSN: 1939-4675

Awwad BSAL (2024) Governance with relationship between entrepreneurship and economic growth in Palestine. Int J Law Manage 66(2):259–287

Awwad BS, Alkababji MW, Zidan S (2023) Governance and social responsibility: Palestinian investment fund. Bus Econ Res 13(4):185–215

Awwad B, El Khoury R (2024) Information technology governance and bank performance: Evidence from Palestine. J Decis Syst 33(2):311–334

Awwad BS (ed) (2025) Sustainability in light of governance and artificial intelligence applications. Emerald Group Publishing.

Bollier D (2007) A new politics of the commons. Renewal (Lond, Engl) 15(4):10–16

Bollier D (2020) Hacking the law to open up zones for commoning. The great awakening: new modes of life amidst capitalist ruins, pp 223–246

Doughnut Economics Action Lab (2024) About doughnut economics: principles of practice [Web page]. https://doughnuteconomics.org/about-doughnut-economics#principles-of-practice

Fanning AL, O'Neill DW, Hickel J, Roux N (2022) The social shortfall and ecological overshoot of nations. Nat Sustain 5(1):26–36

Hunter D (2019) SDLP after 20: sustainable development in the anthropocene. Sustain Dev. L. Pol'y 20:4

Jaber R, Burbar M (2023) Digital transformation/Ramallah municipality. In: Artificial intelligence (AI) and finance. Springer Nature Switzerland, Cham, pp 473–483

Khoday K (2024) Accelerating the green transition: socioecological systems and the future of development: development futures series no 81

Khoffash S (2024) The challenges and opportunities of green financing in Palestine from the perspective of banks. J Infrastruct, Policy Dev 08(02):1–20. https://doi.org/10.24294/jipd.v8i12.8800

McAlpine CA, Seabrook LM, Ryan JG, Feeney BJ, Ripple WJ, Ehrlich AH, Ehrlich PR (2015) Transformational change: creating a safe operating space for humanity. Ecol Soc 20(1)

Raworth K (2018) Doughnut economics: seven ways to think like a 21st century economist. Chelsea Green Publishing

Rockström J, Steffen W, Noone K, Persson Å, Chapin FS, Lambin EF et al (2009) A safe operating space for humanity. Nature 461(7263):472–475

Schneider F, Kallis G, Martinez-Alier J (2010) Crisis or opportunity? Economic degrowth for social equity and ecological sustainability. Introduction to this special issue. J Cleaner Prod 18(6):511–518

SwitchMed (2022) 13 new environmental and social projects in Palestine starting the incubation phase (Web Page). https://switchmed.eu/news/projects-palestine-incubation-phase/

Trindade EP, Hinnig MPF, da Costa EM, Marques JS, Bastos RC, Yigitcanlar T (2017) Sustainable development of smart cities: a systematic review of the literature. J Open Innov: Technol, Market, Complex 3(3):1–14

Turner RA, Wills J (2022) Downscaling doughnut economics for sustainability governance. Curr Opin Environ Sustain 56:101180

United Nations (2023) United Nations sustainable development cooperation framework: state of Palestine 2023–2025 (Report). https://palestine.un.org/sites/default/files/2023-03/UNSDCF%202023-2025_Final.pdf

United Nations Country Team in the Occupied Palestinian Territory, United Nations common country analysis for the occupied palestinian territory (Report, November 2022) https://palestine.un.org/sites/default/files/2023-08/United%20Nations%20Common%20Country%20Analysis%20for%20the%20Occupied%20Palestinian%20Territory.pdf

Urban Agenda Platform (n.d.) WeRamallah smart city initiative (Report). https://www.urbanagendaplatform.org/best-practice/weramallah-smart-city-initiative

World Bank Group (2021) Palestinian cities get a boost through integrated urban planning and innovation (Feature Story). https://www.worldbank.org/en/news/feature/2021/10/14/palestinian-cities-get-a-boost-through-integrated-urban-planning-and-innovation

Yigitcanlar (2006) Australian local governments practice and prospects with online planning. J Urban Reg Inf Syst Assoc 18(2)

The Circular Economy: Business Models for a Regenerative Future

7

George N. Ike and Dike U. Ike

Abstract

The circular economy (CE) has emerged as a transformative paradigm for reshaping global production and consumption patterns toward regenerative, restorative systems. By integrating closed-loop production, waste-to-resource strategies, and innovative value propositions such as product-as-a-service, circular business models decouple economic growth from resource depletion. This article provides a comprehensive analysis of circular business models across multiple industries since 2010, synthesizing peer-reviewed literature, industry reports, and case studies. We explore theoretical underpinnings, implementation pathways, facilitators and obstacles, culminating in actionable recommendations for scholars, policymakers, and practitioners aiming to foster sustainable, resilient economies.

Keywords

Circular economy • Regenerative business models • Closed-loop production • Waste-to-resource • Product-as-a-service • ReSOLVE • Sustainability • Circular business model archetypes

G. N. Ike (✉)
Girne American University, Girne via Mersin 10, Turkey
e-mail: georgeike@gau.edu.tr

Research Center of Energy Economics, Azerbaijan State University of Economics, Baku, Azerbaijan

D. U. Ike
Ted Rogers School of Management, Toronto Metropolitan University, Toronto, ON, Canada

B. Awwad and R. A. Awwad (eds.), *Technology and Innovation Towards a Future of Business Sustainability*, Contributions to Environmental Sciences & Innovative Business Technology, https://doi.org/10.1007/978-981-95-6469-9_7

7.1 Introduction

The traditional linear economy, which is characterized by extraction, production, consumption, and discarding, is an unsustainable system which exerts severe pressure on natural resources, contributes to environmental pollution, and accelerates climate change (Geissdoerfer et al. 2020). In contrast, the circular economy (CE) restores and regenerates systems by keeping products, components, and materials in continuous circulation, and by returning biological nutrients safely to the biosphere (Ellen MacArthur Foundation 2019). CE's three core principles, which include the elimination of waste and pollution, the circulation of products and materials, and the regeneration of natural systems, provide a blueprint for the reimagination and reinvention of value creation (Stahel 2020).

In an era of increasing resource scarcity, volatile commodity prices, and stringent environmental regulations, businesses must adopt novel strategies to enhance resilience and competitive advantage. Circular business models (CBMs) transform companies from mere product providers to stewards of material value, thus unlocking new revenue streams through services, leasing, remanufacturing, and advanced recycling (Bocken et al. 2016).

In this chapter, an attempt will be made to ascertain how CE principles are incorporated into business models in order to make them more viable from a sustainability standpoint. The study will attempt to analyze sector-specific CBM archetypes.

The objectives of this chapter are:

- Synthesize theoretical foundations of CE and delineate key components of CBMs.
- Analyze key literature on the theory and empirics of CBM
- Document sector-specific closed-loop and waste-to-resource practices with quantitative metrics.
- Evaluate and classify CBM archetypes using the ReSOLVE framework and industry typologies.
- Assess policy, technological, financial, and organizational factors influencing CBM scalability.

7.2 Conceptualizing Circular Business Models

Circular business models (CBMs) represent a paradigm shift from linear product transactions to cyclical use and recovery systems. Geissdoerfer et al. (2020) conceptualize CBMs as mechanisms that drive restorative processes through design and innovation, emphasizing longevity, resource recovery, and system regeneration. Suárez-Eiroa et al. (2019) complement this by proposing seven operational principles which are: adjusting inputs to the system to regeneration rates, adjusting outputs from the system to absorption rates, closing the system, maintaining the value of resources within the system, reducing the system's size, designing for CE, and educating for CE. Bocken et al. (2016) identify three business model categories which include *slowing*, *closing*, and *narrowing* resource loops. These business models each address strategies for resource recovery, product life extension, and resource efficiency, respectively, each delivering distinct sustainability outcomes.

The Ellen MacArthur Foundation's **ReSOLVE** framework categorizes CBM strategies into six domains:

- **Regenerate**: Shift to renewable or bio-based inputs (e.g., bio-composite packaging).
- **Share**: Promote shared usage models (e.g., clothing rental services).
- **Optimize**: Enhance resource productivity through predictive maintenance.
- **Loop**: Enable material loops via recycling and remanufacturing.
- **Virtualize**: Replace physical products with digital equivalents (e.g., streaming media).
- **Exchange**: Swap materials for advanced alternatives (e.g., bioplastics).

Accenture refines these into five scalable archetypes: circular supply chain, recovery and recycling, extending product life, sharing platforms, and product-as-a-service. Their 2019 consumer pulse survey revealed that 72% of consumers bought more environmentally friendly products in the last 5 years preceding the survey. Also, 81% pledged to buy more environmentally friendly products in the next 5 years after the survey.

7.3 Methodology

This study conducted a systematic literature review. Web sources were searched for peer-reviewed articles (2010–2025) using keywords: "circular business model," "closed-loop production," "product-as-a-service," and "waste valorization." Grey literature from the Ellen MacArthur Foundation was also reviewed. The article timeline (2010–2025) was predicated on the need to get state-of-the-art insights on CBM. Based on the review, real-world case studies were also reviewed to ascertain if real-world outcomes aligned with empirical research.

Inclusion criteria:

- Empirical studies reporting quantitative or qualitative outcomes of CBM implementations.
- Industry reports detailing pilot projects or mature programs.
- Policy analyses evaluating CE-related regulations and incentives.

Exclusion criteria:

- Articles published before 2010.

Industry case studies were selected to capture each ReSOLVE domain and to ensure balanced sectoral representation.

7.4 Results and Discussion

Table 7.1 summarizes leading CBM insights and typologies from major academic sources between 2010 and 2021, highlighting convergences and divergences in focus areas.

7.4.1 Closed-Loop and Narrowing-Loop Production Mechanisms

Closed-loop production ensures that end-of-life materials re-enter manufacturing processes, reducing raw material extraction. In textiles, Sandin and Peters (2021) compare mechanical versus chemical and thermal fiber recycling. They show that of the 15–20% of disposed textiles that are collected in Europe, 50% are reused while the other 50% is recycled.

Emerging practices in electronics include modular design, which allows for component-level replacement and refurbishment, significantly extending product lifespans (Thompson et al. 2020).

Narrowing loop production, on the other hand, entails resource efficiency, which involves the minimization of material and energy inputs throughout the product lifecycle. Bristol-Myers Squibb, a pharmaceutical firm, successfully reduced the Process Mass Intensity metric of 9 chemical compounds by over 60% through clinical trials. Their efforts have led to the elimination of about 14 million metric tons of waste, reducing both waste disposal costs and CO_2 emissions.

7.4.2 Innovative waste-to-Resource Pathways

Beyond traditional recycling, CE literature spotlights valorization techniques that transform waste into higher-value outputs. Orlando et al. (2023) identify pyrolysis for mixed plastics, enzymatic depolymerization for PET, and advanced composting for organics as key breakthroughs. In agro-industrial contexts, Anaerobic Digestion (AD) not only produces renewable energy but also generates nutrient-rich biofertilizers. Veolia's AD plants convert 14 million tons of organic waste into 1.2 million MWh of electricity annually, sequestering carbon and mitigating landfill methane emissions.

Similarly, a study by Cholake et al. (2018) has discovered a novel way of transforming Electric Arc Furnace (EAF) dust, a steel by-product, into value-added building materials. This can also mitigate significant disposal costs because EAF dust possesses high concentrations of heavy metals, which pose significant environmental risks when being disposed.

Table 7.1 Summary of key literature on circular economy business models

Author(s)	Year	Focus area	Business model typology/ framework	Key insights	Application sectors
Bocken et al	2014	Sustainable business model archetypes	8 archetypes: maximize material productivity, create value from waste, encourage sufficiency, etc.	Introduces core CE business strategies with a sustainability lens	Cross-sector: manufacturing, services, energy
Urbinati et al	2017	Circular economy innovation	4 business model modifications: reverse supply chain activities, pay-per-use approach, higher company to customer cooperation, payment for use, and results	Classifies CE business models and analyzes the internal capabilities required	Manufacturing, consumer goods
Ellen MacArthur foundation	2021	Systems-level framework	ReSOLVE: regenerate, share, optimize, loop, virtualize, exchange	Holistic CE transition strategy combining business and policy	All sectors (global applicability)
Lewandowski	2016	Conceptual framework	CE business model canvas	Extends the traditional business model canvas to include circular dimensions	Business planning, entrepreneurship
Kirchherr et al	2017	CE concept definitions	Review of 114 definitions	Provides definitional clarity and identifies gaps in CE literature	Theoretical, policy, and research relevance
Bocken et al	2016	Product design and business model	Circular design principles: design for longevity, modularity, repairability	Emphasizes product-service synergies and design as an enabler	Design-intensive industries (electronics, appliances)

(continued)

Table 7.1 (continued)

Author(s)	Year	Focus area	Business model typology/ framework	Key insights	Application sectors
Jabbour et al	2018	Industry 4.0 and CE	Integration roadmap: digital tech + CE	Proposes digital enablers for CE (IoT, blockchain, AI)	Industrial manufacturing, logistics
Julianelli et al	2020	Reverse logistics	Taxonomy of CE-enabling logistics practices	Identifies key logistics capabilities to support CBMs	Waste management, retail, e-commerce
Stahel	2020	Resource productivity	Performance economy	Advocates for extended use, service-based models, closed loops	Engineering, urban systems

7.4.3 Sectoral Closed-Loop Production Case Studies

Textiles and Fashion

H&M's Garment Collecting Initiative. Collected over 170,000 tons of textiles globally for reuse and recycling between 2013 and the end of 2023 (H&M 2023). This program aims to prioritize reuse and recycling over waste.

Eileen fisher's Renew Program. Successfully integrates returned garments into resale, repair, and upcycling initiatives. Since 2009, the program has collected over 2 million garments, with approximately half being resold, donated, or recycled.

Patagonia's Worn Wear Program. Operates a robust platform where customers can trade in used garments for store credit, which can be used to purchase other refurbished Patagonia items. This promotes garment longevity and recirculation.

Electronics and Consumer Goods

Nokia's Device Recycling. Nokia participates in global take-back and recycling schemes. Nokia emphasizes design for longevity and recyclability, supporting industry-wide efforts for higher recovery rates (Nokia 2023).

HP's Planet Partners Program. A longstanding leader in cartridge recycling. As of 2023, HP reported using over 50% recycled plastic content in new Original HP cartridges and has collected over 1 billion cartridges cumulatively since 1991 for recycling (HP 2023).

Automotive and Mobility

BMW Group Recycling & Dismantling. BMW operates certified recycling centers and adheres to strict EU End-of-Life Vehicle (ELV) regulations, consistently achieving high recovery/reuse rates for materials from ELVs across its network.

Table 7.2 Diversified CBM archetype outcomes

Archetype	Key metrics	Examples
Closed-loop supply	50–85% material recovery; −15–30% CO_2	Apple: recycling rare earths (2025); reduction of CO_2 via recycling (2023)
Resource recovery	70–85% avg. recycling; ≤ 10% revenue	TerraCycle: recovery of complex waste (2023 impact report)
Product life extension	+ 10–20% lifespan; 10–20% utilization	Rolls-Royce "Power-by-the-hour." Payment of engine maintenance costs to the producer based on per hourly use rates of the purchaser
Sharing platforms	+ 25–35% utilization; −15–20% emissions	Airbnb short-term rentals
Product-as-a-service	40–70% return rate;	Philips LaaS (Light as a service) subscription-based lighting

Tesla & Redwood Materials Partnership. Tesla collaborates with Redwood Materials on lithium-ion battery recycling. Redwood Materials publicly reports achieving recovery rates of over 95% for nickel, cobalt, and copper, and over 80% for lithium from battery materials (Redwood 2024).

7.4.4 Advanced Waste Valorization Pathways

Organic Waste to Energy
Veolia Anaerobic Digestion (AD). Veolia is a major global operator of AD facilities. In 2023, Veolia treated 63 million metric tons of waste globally, including significant organic fractions, producing renewable energy equivalent to the needs of over 500,000 homes and avoiding substantial CO_2 emissions (Veolia 2023).

Industrial By-Product Utilization
Bristol-Myers Squibb Solvent Recycling. BMS implements advanced solvent recovery and recycling technologies across its network, significantly reducing hazardous waste generation and disposal costs. The company highlights continuous improvement in waste reduction and resource efficiency as core sustainability goals (Bristol-Myers Squibb 2023) Table 7.2 outlines some examples of diversified CBM archetypes.

7.4.5 Enabling Policy and Digital Infrastructure

Policy instruments like Extended Producer Responsibility (EPR), Eco-design for Sustainable Products Regulation (ESPR), and green public procurement are important facilitators. The EU's Circular Economy Action Plan has established binding recycled content standards for key products while mobilizing substantial public

R&D funding. In Asia, China's Circular Economy Promotion Law has also driven marked improvements in municipal recycling rates nationwide.

Digital technologies like blockchain, IoT, and AI can facilitate product traceability and lifecycle analysis. Studies indicate blockchain-based product passports significantly reduce material misclassification errors, while AI-driven predictive maintenance platforms notably extend operational uptime in manufacturing, exemplifying the Optimize archetype.

7.4.6 Facilitators and Obstacles

Facilitators

Regulatory Frameworks. EPR schemes across OECD countries have dramatically increased electronics recycling rates.

Financial Instruments. Green bonds and circularity-linked loans have been able to mobilize significant capital for global CE projects.

Digital Platforms. Material passports demonstrate meaningful reductions in sorting errors, enhancing recyclable feedstock quality.

Awareness. Increasing awareness of CE and CBM concepts can induce a positive behavioral change in consumers and firms alike. The environmental awareness of EU citizens went from about 27% in 2007 to close to 60% in 2024. In Fig. 7.1, Google search trends on the search items "Circular Economy" (CE) and "Circular Business Model" (CBM) show an increasing trend, with CE having a much earlier search activity than CBM because of its earlier appearance in the literature.

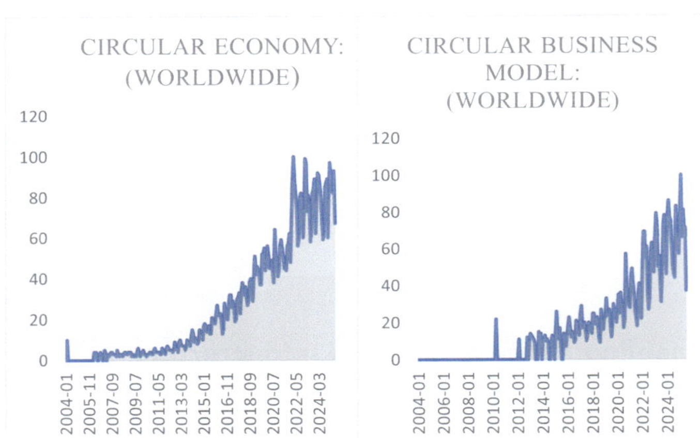

Fig. 7.1 Google trends search interest in circular economy and circular business model from 2004 to 2025 *Source* Statistical analysis outputs

Obstacles

Capital Expenditure. SMEs face disproportionately high upfront costs for circular infrastructure, with protracted payback periods.

Skills Gap. A widespread lack of specialized expertise hampers CE adoption, necessitating the development of a designated workforce.

Cultural Resistance. Entrenched short-term profit metrics can actively constrain the transition to circular business models because of investment costs.

7.5 Conclusion

This article demonstrates that the circular economy offers a robust framework for embedding regenerative principles into business models across sectors. By advancing closed-loop production, waste valorization, and innovative service-based archetypes, firms can realize substantial environmental and economic gains. Policy support, financial innovation, digital traceability, and capacity building are pivotal enablers, while cost barriers, skill shortages, and cultural inertia remain challenges. Future research should explore the longitudinal impacts of CBM adoption and develop standardized metrics for circularity performance. Realizing a truly regenerative future will require concerted collaboration among academia, industry, and governments to scale circular models and ensure resilient, sustainable growth.

7.5.1 Policy Recommendations

Government and concerned stakeholders should implement targeted subsidies for reverse logistics and recycling infrastructure. They should also promote cross-sector CE clusters to facilitate industrial symbiosis and integrate CE modules into vocational and business education curricula. Finally, there should be a standardization of digital product passport frameworks to boost traceability.

References

Bocken NM, Short SW, Rana P, Evans S (2014) A literature and practice review to develop sustainable business model archetypes. J Clean Prod 65:42–56

Bocken N, de Pauw I, Bakker C, van der Grinten B (2016) Product design and business model strategies for a circular economy. J Ind Prod Eng 33(5):308–320

Bristol-Myers Squibb (2023) ESG Report

Cholake ST, Farzana R, Numata T, Sahajwalla V (2018) Transforming electric arc furnace waste into value added building products. J Clean Prod 171:1128–1139

Danish Energy Agency (2021) Biogas map Denmark 2021. DEA, Copenhagen

Eileen Fisher (2023) Renew impact Report

Ellen MacArthur Foundation (2019) Completing the picture: how the circular economy tackles climate change. Cowes, UK

European Commission (2022) EU circular economy action plan. EC, Brussels

Geissdoerfer M, Savaget P, Bocken NMP, Hultink EJ (2020) The circular economy—a new sustainability paradigm? J Clean Prod 243:118609
H&M Group (2023) Sustainability report
Hollander M, Koot H, Vermeulen W (2018) Barriers to the adoption of circular economy practices in SMEs: a systematic literature review. J Clean Prod 195:295–307
HP (2023) Sustainable impact report
Julianelli V, Caiado RGG, Scavarda LF, Cruz SPMF (2020) Interplay between reverse logistics and circular economy: critical success factors-based taxonomy and framework. Resour Conserv Recycl 158:104784
Kirchherr J, Reike D, Hekkert M (2017) Conceptualizing the circular economy: an analysis of 114 definitions. Resour Conserv Recycl 127:221–232
Lewandowski M (2016) Designing the business models for circular economy—towards the conceptual framework. Sustainability 8(1):43
Lopes de Sousa Jabbour AB, Jabbour CJC, Godinho Filho M, Roubaud D (2018) Industry 4.0 and the circular economy: a proposed research agenda and original roadmap for sustainable operations. Ann Oper Res 270(1):273–286
Nokia (2023) People and planet report
OECD (2023) Extended producer responsibility: policy approaches and implementation. OECD, Paris
Orlando M, Molla G, Castellani P, Pirillo V, Torretta V, Ferronato N (2023) Microbial enzyme biotechnology to reach plastic waste circularity: current status, problems and perspectives. Int J Mol Sci 24(4):3877
Patagonia (2023) Worn wear. [Internet]. Available from: https://wornwear.patagonia.com/
Redwood materials (2024) Impact report
Sandin G, Peters GM (2021) Environmental impact of textile reuse and recycling—a review. J Ind Ecol 25(2):262–279
Stahel WR (2020) The history of the circular economy. In: The circular economy in the European Union: an interim review, pp 7–19
Suárez-Eiroa B, Fernández E, Méndez-Martínez G, Soto-Oñate D (2019) Operational principles of circular economy for sustainable development: linking theory and practice. J Clean Prod 214:952–961
Thompson DL, Hartley JM, Lambert SM, Shiref M, Harper GD, Kendrick E et al (2020) The importance of design in lithium ion battery recycling—a critical review. Green Chem 22(22):7585–7603
Urbinati A, Chiaroni D, Chiesa V (2017) Towards a new taxonomy of circular economy business models. J Clean Prod 168:487–498
Veolia (2023) Universal registration document
World Economic Forum (2021) Global risks report 2021. WEF, Geneva

From Linear to Circular: Supply Chain Business Models for a Regenerative Economy

8

Najwan Ibrahim Jadallah, Farah Essa Dudin, and Bahaa Awwad

Abstract

The move from linear to circular supply chains is at the heart of regenerative economic models in their transformation to provide sustainability, resilience, and value retention. Circular design could be made by reverse logistics and digital enhancement techniques with proper control through blockchain, IoT, and AI technologies. This would allow for optimal resource efficiency with waste minimization through opening up new product lifecycles. Product-as-a-service models that further include refurbishing and collaborative consumption are changing the paradigms of use and value attached to products; they are making environmental care profitable. Such cross-sectoral and regional strategic collaborations, supported by enabling environments and financial mechanisms, actually make it possible to go beyond all systematic barriers in scaling circular solutions. These will bring about transparency, traceability, and real-time decision support for digitally enhanced consumers, as well as behavioral incentives for consumer engagement in recovery and reuse activities. As long as sustainability remains at the core of business operations, organizations implementing circular strategies gain a competitive advantage, resilient supply chains, and contribute to ecological balance in the long run. Circular supply chains are not optional; they constitute the future of responsible, high-impact, inclusive economic transformation.

N. I. Jadallah (✉) · F. E. Dudin
Palestine Technical University—Kadoorie, Ramallah, Palestine
e-mail: najwan.abujadalla@ptuk.edu.ps

F. E. Dudin
e-mail: Farah.dudin@ptuk.edu.ps

B. Awwad
College of Business and Finance, Ahlia University, Manama, Bahrain
e-mail: bawwad@ahlia.edu.bh

B. Awwad and R. A. Awwad (eds.), *Technology and Innovation Towards a Future of Business Sustainability*, Contributions to Environmental Sciences & Innovative Business Technology, https://doi.org/10.1007/978-981-95-6469-9_8

Keywords

Circular economy • Regenerative supply chain • Sustainability innovation • Reverse logistics • Product-as-a-service

8.1 Introduction: Redefining the Path of Supply Chains

The shift from linear "take-make-dispose" models to circular supply chains is critical to building regenerative economies. This change goes beyond sustainability by incorporating methods that improve fairness, environmental resilience, and long-term sustainability. Through tactics like inclusive growth and local sourcing, businesses are now taking into account their effects on ecosystems and communities in addition to profit metrics.

Effective and responsible circular supply chains are supported by cutting-edge technologies like blockchain for transparency and data analytics for performance enhancement. While cross-sector cooperation and policy tools like Extended Producer Responsibility (EPR) increase accountability and innovation across the product lifecycle, closed-loop systems lessen dependency on raw materials.

As consumer demand for sustainable products rises, circular models also create new market opportunities. This change is made possible by education and workforce development, which give people the tools they need to support circular operations. This chapter examines the models and strategies shaping the regenerative supply chains of the future.

8.2 The Foundational Elements of a Circular Supply Chain

Creating a successful circular supply chain requires close collaboration among stakeholders and the strategic use of digital tools to optimize resource management throughout a product's lifecycle. This holistic approach supports the transition to a truly regenerative economy and relies on systematic thinking about design, supply, and end-of-life recovery strategies (Alcayaga and Hansen 2025).

8.2.1 Circular Design and Sourcing

Circular design and sourcing are essential components of a circular supply chain, starting with the use of sustainable and renewable inputs. Informed consumers are increasingly demanding environmentally friendly products, a trend supported by take-back schemes and digital technologies such as blockchain, which enhance traceability and transparency across supply networks (Farazi 2024). This supports trust and accountability in the sourcing process, which is critical for the adoption of circular practices.

Designing for durability, repair, and modular assembly is critical. Regulatory mechanisms, including Extended Producer Responsibility (EPR) and Eco-design Directives, require producers to consider product performance throughout its lifecycle (Barrie et al. 2022). These actions not only encourage creativity but also make it possible to use materials that are simpler to recycle, disassemble, or refurbish. Recent studies show that companies that implement these modular strategies improve value retention and reduce their environmental footprint (Jum'a et al. 2024).

Disassembly and end-of-life planning reimagine product development by integrating circularity from the outset. Products designed for disassembly facilitate efficient component separation, improve material recovery, reduce waste, and support closed-loop manufacturing systems. Digital product passports and IoT-enabled traceability enhance product lifecycle visibility, particularly during the recovery and remanufacturing phases.

Another important factor in advancing circularity is consumer engagement. Strategies that encourage customer returns, subscription models, and product leasing help ensure resource recycling, extend product life, and reduce reliance on raw materials. Collaborative ecosystems, especially those involving cross-sector partnerships and industry alliances, facilitate knowledge exchange and infrastructure sharing, which is essential for scaling up circular resource and material recovery initiatives (Solomon et al. 2024).

8.2.2 Reverse Logistics and Collection Networks

Reverse logistics and collection networks are essential for closing the loop by reintegrating end-of-life products into production systems, preserving their material value, and reducing environmental impact. Strong infrastructure, cooperation from stakeholders, and supportive regulations are essential to their success. Efficient collection networks—characterized by accessible drop-off points and responsive logistics—maximize material recovery and minimize landfill use (Daramola et al. 2023).

IoT-enabled technologies improve visibility across the collection, transportation, and remanufacturing stages, supporting predictive routing and condition-based sorting to improve scalability. However, challenges remain in emerging economies due to limited institutional support and consumer skepticism (Onukwulu et al. 2025).

Product take-back programs rely on transparent returns and incentives, while consumer engagement is critical. Data-driven feedback also improves waste collection strategies and builds trust in circular systems (Alcayaga and Hansen 2025; Daramola et al. 2023).

Digitization and data are essential components for enabling circular supply chains by improving traceability, predictability, and coordination across complex networks. Stakeholder trust is increased, inefficiencies are decreased, and decision-making is improved by these technologies.

The combination of blockchain technology and the Internet of Things (IoT) improves supply chain transparency and traceability. Blockchain provides immutable records of product journeys, supporting regulatory compliance and supplier accountability (Rejeb et al. 2025). When combined with the Internet of Things, companies gain real-time data on the location and condition of materials, enhancing transparency and consumer trust (Haar 2022).

Improving backflows and predicting product end-of-life is enhanced by digital tools such as smart sensors and analytics. These technologies enable tracking of product consumption, expedite timely returns, and simplify logistics processes to reduce recovery costs (Alcayaga and Hansen 2025; Butt et al. 2023).

Real-time feedback systems, demand forecasting, and predictive analytics are made possible by data-driven cooperation throughout the supply chain. Platforms powered by artificial intelligence and machine learning help manage resources dynamically and improve sustainability reporting. The integration of digital infrastructures also supports new circular models such as product-as-a-service, enhancing resource efficiency and long-term value. Adopting this digital foundation is a strategic imperative for resilient and regenerative supply chains (Bressanelli et al. 2024).

8.3 Core Supply Chain Business Models for Circularity

The development of circular supply chains increasingly relies on collaborative ecosystems that unite businesses, governments, and civil society to share resources, exchange knowledge, and coordinate value recovery. These partnerships address system fragmentation and are essential for promoting circularity (Onukwulu et al. 2025).

For example, municipal collaborations promote community recycling programs, improving accessibility and reducing landfill waste. In resource-constrained areas, public–private partnerships contribute to the establishment of recycling centers, logistics services, and educational campaigns. These initiatives result in shared logistics networks that improve circular efficiency, cut costs, and reduce emissions. Key circular business models support this transformation:

- Product-as-a-Service (PaaS): Extends product lifespan and increases return rates, particularly in the electronics and textile sectors (Hidalgo-Crespo et al. 2024).
- Resource recovery: Focuses on material recycling and component collection, enabled by efficient reverse logistics and data systems.
- Circular inputs: Focuses on renewable or recycled materials to close production loops (De Angelis 2024).

Digital platforms enable these models by supporting real-time traceability, material transparency, and secondary market coordination (Alcayaga and Hansen 2025). They also provide data for continuous improvement and stakeholder alignment. The foundation of regenerative supply chains is made up of digital innovation and collaboration, which fundamentally alters the way value is produced, delivered, and recovered.

- Product-as-a-Service (PaaS) Supply Chain

The product-as-a-service (PaaS) model transforms consumption by moving from ownership to access, stimulating shared responsibility throughout the product lifecycle. This approach encourages the return, reuse, and upgrade of products, aligning economic value with environmental goals.

Successful PaaS implementation requires understanding consumer willingness to engage in circular practices. Companies use data analytics to design services, improve user experiences, and foster loyalty through environmentally friendly value propositions (Ezekwu 2025). Feedback loops from digital platforms enable smarter inventory planning and environmentally conscious product redesign (Onukwulu et al. 2022; Alcayaga and Hansen 2025).

PaaS restructures supply chains by incorporating maintenance, repair, and upgrades. Predictive maintenance powered by the Internet of Things and machine learning extends product life and reduces waste (Khompatraporn 2021). Modular designs support repairability and customization, reducing reliance on raw inputs and promoting circularity (Kesavaraj 2024).

Scalability relies on the efficiency of reverse logistics. Real-time tracking, smart labeling, and geolocation improve product returns and sorting (Daramola et al. 2023; Alkayaga and Hansen 2025). To cut waste and carbon emissions, businesses like Xerox and Michelin have implemented closed-loop platform-as-a-service (PaaS) models.

Consumer education remains critical. Campaigns and case studies—such as Philips' Circular Lighting program—highlight the environmental benefits of take-back schemes and foster engagement (Onukwulu et al. 2025). PaaS aligns technology, logistics, and consumer behavior, positioning companies to lead the way in building resilient and sustainable supply chains.

8.3.1 Refurbishment, Repair, and Remanufacturing Supply Chains

Integrating refurbishment, repair, and remanufacturing into supply chains enhances sustainability and resource efficiency, but poses strategic challenges. Companies must invest in modular design, skilled labor, and diagnostic technologies to ensure remanufactured products meet "new" standards while remaining cost-effective (Barrie et al. 2022; De Angelis 2024; Fraccascia et al. 2019).

Consumer skepticism remains a barrier to expansion. Educational campaigns coupled with certifications and warranties help change perceptions, as companies like Caterpillar and Bosch have demonstrated (Chen et al. 2020). Implementing these models requires managing two-way flows and fostering partnerships between manufacturers, retailers, and recyclers to support coordinated recovery (Kesavaraj 2024).

Digital platforms play a pivotal role by enabling immediate access to parts inventories and standardizing reverse logistics processes across regions (Alcayaga and Hansen 2025). Sourcing components through product returns or third-party centers reduces reliance on raw materials and supports circular value chains.

Effective monitoring relies on key performance indicators (KPIs), such as recovery rates, energy consumption, and delivery time, which provide transparency and support environmental, social, and governance (ESG) accountability (Aljamal et al. 2024; Cao and Tao 2023). Consumer engagement is also critical, as tools such as sustainability dashboards and lifecycle calculators boost participation in recovery programs and raise awareness (Onukwulu et al. 2022).

By incorporating these models into supply chain strategies, companies enhance their resilience, reduce waste, and strengthen their competitive position in sustainability-focused markets.

8.3.2 Resource Recovery and Recycling Supply Chains

Establishing efficient supply chains for resource recovery and recycling is essential to a circular economy, reducing waste and environmental damage and treating waste as a strategic input (De Angelis 2024; Onukwulu et al. 2022). Artificial intelligence and Internet of Things technologies now enable precise sorting and automated material recovery, significantly improving recycling efficiency for high-impact industries such as plastics and electronics (Fraccascia et al. 2019; Singh 2022).

Local recycling centers endorse sustainability by engaging communities, reducing transportation emissions, and creating jobs in sorting and repair (GIZ 2023). This infrastructure, when supported by policies and logistics, helps decentralize recovery efforts and build supply chain resilience (Alcayaga and Hansen 2025).

Public–private partnerships and digital platforms are vital for securing secondary materials and coordinating shared logistics services. These networks support compliance and traceability while enabling material markets and immediate collaboration (Taddei et al. 2024). Specialized processing facilities, supported by investment and innovative financing such as green bonds, are essential for scaling up closed-loop systems (Onukwulu et al. 2022).

Chemical recycling technologies—such as pyrolysis and gasification—enable the conversion of complex wastes into high-quality raw materials, addressing materials previously considered non-recyclable and reducing reliance on virgin inputs (Ministry of Local Government 2023).

Finally, designing for recyclability and incorporating lifecycle thinking into procurement processes ensures that recovered materials are effectively reintegrated. This strengthens feedback loops, enhances accountability, and unlocks long-term environmental and economic value (Ingaldi and Ulwitz 2024).

8.3.3 Collaborative Consumption Supply Chains

Collaborative consumption supply chains offer a transformative model that reduces resource use, extends product life cycles, and encourages community sustainability through shared access rather than ownership (Marsala et al. 2022; Shahvaroughi et al. 2024). These models, which represent shifting consumer values, encourage cost savings, ethical consumption, and the development of green jobs.

Platforms that facilitate tool libraries, fashion rentals, or co-working spaces improve asset utilization and reduce demand for new goods, encouraging sustainable behaviors and social cohesion (Arrigo 2021; Surahman and Hariyadi 2021). Inventory management, logistics, and maintenance of shared-use products present distinct supply chain challenges, requiring flexible systems that respond to fluctuating usage patterns (Hernandez Marquina et al. 2024; Kanwal et al. 2024).

Policy support is critical. Incentives such as tax breaks, infrastructure funding, and flexible zoning foster the adoption of sharing platforms, while clear legal frameworks help address liability and privacy issues (Shahvaroughi et al. 2024). Trust is equally important, as platforms must implement verification systems, maintenance standards, and peer review mechanisms to ensure product quality and user trust (Wen and Siqin 2020).

Digital technologies underpin these systems. Blockchain and the Internet of Things provide transparent records and enable real-time tracking and predictive maintenance. Data analytics support demand forecasting and efficient asset allocation (Onukwulu et al. 2022; Razia and Awwad 2022; Rejeb et al. 2025). Collaborative consumption redefines supply chains around shared value, resilience, and environmental responsibility, providing a scalable path toward more inclusive and regenerative economies.

8.4 Strategic and Operational Shifts for Supply Chain Managers

Supply chain managers transitioning to a circular economy must prioritize strategic partnerships. As a way to spur innovation, make resource sharing easier, and create a strong ecosystem that comprises suppliers, clients, and even rival businesses, these partnerships are crucial. Through expanding horizons and creating innovative solutions, these collaborations improve operational efficiency, encourage sustainability, minimize waste, maximize resource utilization, and help create a sustainable future.

8.4.1 Shifting from Cost Minimization to Value Maximization and Cultivating Supply Chain Collaboration

The transition to a circular economy requires a fundamental shift in supply chain strategy, from traditional cost minimization to value maximization across economic, environmental, and social dimensions. This shift prompts companies to reevaluate performance metrics, moving beyond price and efficiency to include long-term material retention, carbon footprint reduction, and product lifecycle sustainability (Marsala et al. 2022; Singh 2022). Circular value chains focus on resource loops, reducing reliance on raw materials and creating a regenerative impact through product lifespan extension, reuse, and recovery (Guidelines for Measuring the Circular Economy 2024). When determining the actual cost of a good or service, life cycle costing, or LCC, becomes an essential tool. LCC takes into account downstream waste treatment, upstream material extraction, and externalities like social costs and environmental degradation, in contrast to traditional costing. LCC strengthens supply chain transparency and encourages responsible sourcing when paired with Environmental Product Declarations (EPDs) (Onukwulu et al. 2022). Furthermore, cutting-edge financing methods are becoming more popular for funding advanced recycling technologies, remanufacturing facilities, and digital infrastructure. These include green bonds, sustainability-linked loans, and environmental, social, and governance (ESG) investment tools (De Angelis 2024). Enhancing collaboration across the supply chain is essential to unlocking these opportunities. The Sustainable Development Goals (SDGs) can be aligned with operational capabilities through cross-functional coordination between the design, procurement, logistics, and sustainability departments. Strong frameworks for assessing, tracking, and reporting environmental performance must be implemented by businesses. Reporting frameworks, such as the Global Reporting Initiative (GRI), the Sustainability Accounting Standards Board (SASB), and ISO 14001, support the standardized tracking of environmental and social key performance indicators (Ministry of Local Government 2023; Solomon et al. 2024). This shift also fosters industrial symbiosis—a collaborative model in which one company's waste or by-products become input materials for another. This model enhances local economic ecosystems, facilitates resource efficiency, and lessens dependency on landfills. Circular industrial parks embody this model by clustering companies with complementary resource needs, optimizing shared logistics and utility infrastructure, and reducing emissions (Carlucci et al. 2021; Gabbur 2020). Digital platforms can further support these exchanges by matching supply and demand for secondary materials in real time.

Strategic collaborations extend beyond traditional supply chain partners to include competitors, startups, research institutions, and municipalities, forming innovative ecosystems that accelerate circular innovation for products and processes. Trust-building mechanisms, data-sharing agreements, and shared ownership of sustainability goals foster a culture of shared transparency and accountability (Alcayaga and Hansen 2025; Jum'a et al. 2024). This shift redefines supply chain leadership, rather than focusing solely on lean operations and margin maximization, circular supply chain managers become organizers of regenerative systems who balance value creation, resilience, stakeholder engagement, and long-term sustainability.

8.4.2 Navigating Regulatory and Policy Landscapes and Measuring and Reporting on Circularity

A successful transition to a circular supply chain requires organizations to navigate an evolving and often fragmented regulatory and policy environment. In order to guarantee compliance and foster frameworks that encourage circular innovation, proactive interaction with legislators is essential. Understanding and leveraging Extended Producer Responsibility (EPR) programs enables companies to internalize lifecycle impacts and invest in recovery and recycling systems (De Angelis 2024; Carlucci et al. 2021). Participating in multi-stakeholder policy dialogues and coalitions helps shape supportive regulatory environments while building trust with civil society.

Partnerships between the public and private sectors are becoming more widely acknowledged as useful instruments for expanding circular solutions. By engaging in policy, infrastructure, and logistics systems, these collaborations can address systemic challenges such as waste collection, secondary material standards, and circular procurement (Jum'a et al. 2024). These partnerships have proven particularly effective at the local level, where municipalities are collaborating with companies to pilot circular industrial zones and integrate digital traceability platforms (GIZ 2023).

Environmental certifications and labels are becoming essential instruments for attaining market distinction and regulatory compliance. Labels such as Cradle to Cradle, the EU Ecolabel, and ISO 14001 help companies communicate their sustainability credentials, navigate procurement standards, and access green financing mechanisms (Porwal 2023; Hammarlund et al. 2024). For consumers, these certifications build trust and demonstrate credibility. Businesses that follow third-party verification guidelines will benefit in the marketplace as transparency becomes a differentiator. Businesses must put in place thorough frameworks for measuring and reporting on circularity in order to evaluate and share progress. The cost and production-focused linear metrics of the past need to give way to multifaceted systems that take economic, social, and environmental performance into account. Key performance indicators (KPIs) for circular supply chains include material circularity indicators, lifecycle emissions, waste diversion rates, and stakeholder engagement metrics (Aljamal et al. 2024; Onukwulu et al. 2025). Setting KPIs with the help of stakeholders, including suppliers, local communities, and regulators, improves legitimacy and inclusivity (Nika et al. 2021). Circular measurement is further supported by advanced technologies. Real-time tracking of material flows, product usage, and environmental impacts can be achieved through data analytics, blockchain technology, and Internet of Things (IoT) devices. Machine learning models provide predictive insights into resource needs and product return cycles, enabling companies to better plan inventory, maintenance, and reverse logistics (Alfina et al. 2024). By incorporating user feedback, organizations gain richer data sets that reflect actual performance and align with ESG reporting expectations. Organizations can evaluate progress, locate obstacles, and establish a transparent and accountable culture by implementing efficient measurement and reporting systems. As stakeholders increasingly demand verifiable evidence of sustainability, these capabilities will become essential for maintaining competitiveness and compliance in the global shift toward circularity.

8.5 Challenges and Overcoming Barriers

To make the shift from linear to circular supply chain models, a number of behavioral, technical, and systemic obstacles must be overcome. These barriers, ranging from infrastructure constraints to regulatory fragmentation and consumer inertia, can hinder the scalability of circular practices. Addressing them requires cross-sector collaboration, forward-looking regulatory frameworks, and strategic investments in digital infrastructure and workforce development (Barrie et al. 2022; Carlucci et al. 2021).

For instance, in politically fragile contexts like Palestine, the transition to circular supply chains hinges on effective governance and adaptive entrepreneurship. Political instability and weak institutional capacity often disrupt infrastructure and undermine long-term sustainability planning (Jadallah and Bhatti 2020). To address these challenges, governance frameworks must prioritize transparency, regulatory clarity, and environmental disclosure (Awwad 2022), creating an environment where circular startups and SMEs can thrive. Entrepreneurship, particularly when supported by knowledge-based capabilities and targeted policy incentives, becomes a key driver of localized innovation. Integrating artificial intelligence and digital governance further enhances competitiveness and resilience across sectors (Razia et al. 2024). By aligning circular strategies with national development goals and fostering cross-sector collaboration, Palestine can build regenerative business models that support inclusive growth and ecological sustainability—even amid geopolitical uncertainty.

8.5.1 The Complexity of Reverse Logistics

Reverse logistics is a cornerstone of circular supply chains, but it presents unique operational challenges due to the complexity and diversity of returns. The multidirectional movement of used goods, spare parts, and recycled materials is beyond the capabilities of traditional logistics systems, which are optimized for unidirectional flows. To overcome these shortcomings, technological advancements are crucial, especially in the areas of machine learning, artificial intelligence, and the

Internet of Things. These tools help improve routing, sorting, and condition assessment, reducing costs and improving the speed and accuracy of returns (Bakas et al. 2021; Daramola et al. 2023).

Unpredictable return volumes and the variability in the quality of returned goods further complicate reverse logistics operations. Inconsistent or damaged returns pose challenges to remanufacturing, refurbishment, and recycling operations, requiring advanced diagnostic systems and flexible processing lines. Therefore, circular business models must build adaptable logistics networks capable of absorbing these fluctuations without compromising efficiency or sustainability (De Angelis 2024).

A key factor in the efficiency of reverse logistics systems is consumer behavior. Educational initiatives that raise awareness of responsible waste disposal, return incentives, and environmental impacts significantly increase participation in return programs (Ezekwu 2025). Consumer compliance and returns are improved when circular principles are incorporated into product labels, user manuals, and digital communications (GIZ 2023). Sustainability platforms that provide incentives, including loyalty points or discounts, are more effective at driving user involvement.

Moreover, cross-sector collaborations are essential to the success of reverse logistics Ruzieh et al. (2025). The creation of shared infrastructure and platforms for tracking, sorting, and transporting materials is made possible by cooperation between manufacturers, logistics companies, recyclers, municipalities, and tech firms. These ecosystems reduce the burden on individual actors and enhance collective sustainability impacts (Bakås et al. 2021). With funding from the public and private sectors, regional reverse logistics hubs can improve circular infrastructure, lower emissions from transportation, and generate green jobs. By addressing logistics inefficiencies and leveraging digital and behavioral innovations, organizations can transform reverse logistics from a cost center into a strategic advantage, supporting circular value creation and operational resilience.

8.5.2 Financial and Investment Challenges

The transition to a circular economy has significant financial and investment challenges, particularly in the areas of infrastructure and technology expansion. Adoption of many circular initiatives is hampered by their high initial capital requirements, especially for small and medium-sized businesses (SMEs).To overcome this, public–private partnerships (PPPs) represent a strategic mechanism for pooling resources and sharing risks. These collaborations finance advanced recycling technologies, sorting systems, and digital platforms that simplify reverse logistics and data analytics (Mahjoob et al. 2023; Rajčić et al. 2024).

Financial incentives for businesses that implement circular practices are offered by green finance initiatives, such as sustainability-linked loans, green bonds, and ESG-compliant investment funds. These instruments lower capital costs and attract impact-oriented investors, balancing financial performance with environmental and social outcomes. Governments and development banks also play an incentive role by de-risking investments and providing guarantees or subsidies for circular infrastructure (Onukwulu et al. 2022). Alternative financing models, such as crowdfunding and community finance, have emerged as inclusive mechanisms for circular innovation, particularly in underserved markets (Erp 2022; Ryan and Vladimirova 2023). These strategies encourage local ownership and accountability while making capital more accessible. Furthermore, circular business models, such as leasing, sharing, and subscription services, reshape ownership structures, reducing waste and encouraging the return of products for reuse or refurbishment (Munir and Fausiah 2025). These models ease capital burdens on consumers while stabilizing long-term revenue streams for businesses. These mechanisms help the circular economy grow, reduce risk aversion, and support an inclusive and resilient growth model by redefining finance as a tool for systemic change.

8.5.3 Changing Corporate Culture and Mindset

One of the most persistent barriers to circularity is entrenched linear thinking within organizations. It takes visionary leadership to integrate sustainability into fundamental principles, governance frameworks, and performance rewards in order to transform corporate culture. Leadership that champions environmental responsibility and values long-term sustainability over short-term gains fosters a mindset conducive to innovation and resilience.

An important factor in revolutionizing product development is eco-design training. One. By equipping design teams with circular principles such as modularity, reparability, and disassembly, companies reduce waste and enhance lifecycle management. Embedding these values in research, development, and procurement leads to structurally sustainable products and services (Nousala and Metcalfe 2024).

Workforce development is equally critical. As supply chains become increasingly technologically and operationally complex, employees must acquire technical and soft skills tailored to circular systems. Technical skills include reverse logistics, lifecycle analysis, and IoT-enabled traceability. Soft skills include cross-functional collaboration and stakeholder engagement (Rejeb et al. 2025). Educational programs and internal training on circular economy models, sustainable innovation, and responsible automation empower employees to lead transformation from within (Hobbs 2021).

Businesses that make investments in culture development and upskilling not only become more adaptable but also position themselves as sustainability leaders, encouraging innovation, loyalty, and long-term competitive advantage.

8.6 Conclusion: The Regenerative Supply Chain

The evolution from linear to circular supply chains goes beyond an operational shift; it represents a paradigm shift toward regenerative economies. The supply chain is viewed in this model as a force for social justice, ecological restoration, and economic sustainability rather than as a means of extraction and disposal.

Stakeholder cooperation, cutting-edge technologies, and innovative business models are given top priority in circular supply chains in order to reduce environmental effects and improve resource efficiency. They move beyond transactional scales toward inclusive value creation, integrating environmental management, economic innovation, and social engagement. Regulatory frameworks and education are pivotal in guiding this transformation and ensuring alignment across sectors and regions (De Angelis 2024; Hammarlund et al. 2024).

Closed material loops, regional recovery systems, and online forums that encourage openness and community involvement are all characteristics of regenerative supply chains. These components improve long-term competitiveness, operational resilience, and resilience to external shocks. In particular, consumer engagement and education are key components of success. Raising awareness of the environmental impact of consumption choices fosters responsible behavior and supports product return, reuse, and recycling initiatives (Bargavi et al. 2024).

Gamification technologies, such as point-based recycling systems or digital incentives for take-back programs, improve material recovery rates while enhancing customer loyalty. Social media and mobile platforms create channels for instant engagement, reinforcing sustainability values and establishing ethical and forward-looking companies (Ryan and Vladimirova 2023).

At its core, the future of supply chains lies in their ability to regenerate rather than deplete to create ecosystems where nature, business, and society thrive in harmony. Embracing this vision is not just a business imperative; it is a moral responsibility in the face of global challenges and societal transformation.

References

Alcayaga A, Hansen EG (2025) Smart circular economy as a service business model: an activity system framework and research agenda. R&D Manage 55(2):508–530

Alfina KN, Ratnayake RMC, Wibisono D, Basri MH, Mulyono NB (2024) Prioritizing performance indicators for the circular economy transition in healthcare supply chains. Circular Econ Sustain. https://doi.org/10.1007/s43615-024-00436-9

Aljamal D, Salem A, Khanna N, Hegab H (2024) Towards sustainable manufacturing: a comprehensive analysis of circular economy key performance indicators in the manufacturing industry. Sustain Mater Technol. https://doi.org/10.1016/j.susmat.2024.e00953

Arrigo E (2021) Digital platforms in fashion rental: a business model analysis. J Fashion Market Manage. https://doi.org/10.1108/JFMM-03-2020-0044

Arsawan IWE, Suhartanto D, Koval V et al (2024) Enhancing the circular economy business model towards sustainable business performance: moderating the role of environmental dynamism. J Infrastruct, Policy Dev 8(5):3321. https://doi.org/10.24294/jipd.v8i5.3321

Awwad BS (2022) The role of environmental disclosure in enhancing the market value of industrial and service companies listed on the Palestinian stock exchange. Int J Entrepreneurship 26(2):1–12. ISSN: 1939-4675

Awwad BSAL (2024) Governance with relationship between entrepreneurship and economic growth in Palestine. Int J Law Manage 66(2):259–287. https://doi.org/10.1177/2393957524123456

Awwad BS, Alkababji MW, Zidan S (2023) Governance and social responsibility: Palestinian investment fund. Bus Econ Res 13(4):185–215. https://doi.org/10.5296/ber.v13i4.21353

Bakås O, Tveit SS, Thomassen MK (2021) Reverse logistics for improved circularity in mass customization supply chains. https://doi.org/10.1007/978-3-030-90700-6_118

Bargavi N, Irfana S, Ramana AV, Shankar G, Nagpal P, Dhote S (2024) Circular economy towards sustainable businesses: exploring factors shaping adoption and implementation barriers. Educ Adm: Theory Pract 30(3):813–819

Barrie J, Schröder P, Schneider-Petsinger M, King R, Benton T (2022) The role of international trade in realizing an inclusive circular economy. Chatham house, environment and society programme

Biancolin M (2023) Reverse logistics and circular economy: a literature review. Eur Transp. https://doi.org/10.48295/et.2023.94.7

Brändström J, Jazairy A, Lindgreen ER (2024) Barriers to adopting circular business models: a cross-sectoral analysis. Bus Strateg Environ 33:4331–4350. https://doi.org/10.1002/bse.3653

Bressanelli G, Saccani N, Perona M (2024) Are digital servitization-based circular economy business models sustainable? A systemic what-if simulation model. J Clean Prod 458:142512. https://doi.org/10.1016/j.jclepro.2024.142512

Butt AS, Ali I, Govindan K (2023) The role of reverse logistics in a circular economy for achieving sustainable development goals: a multiple case study of retail firms. Prod Plann Control. https://doi.org/10.1080/09537287.2023.2197851

Cao Z, Tao L (2023) Green finance and economic resilience: Investigating the nexus with natural resources through econometric analysis. Econ Anal Policy. https://doi.org/10.1016/j.eap.2023.09.033

Carlucci S, Lange MA, Santamouris M, Attia S (2021) Eastern Mediterranean and middle east climate change initiative: report of the task force on the built environment

Chen Y, Wang J, Jia X (2020) Refurbished or remanufactured? An experimental study on consumer choice behavior. Front Psychol. https://doi.org/10.3389/FPSYG.2020.00781

Daramola OM, Apeh CE, Basiru JO, Onukwulu EC, Paul PO (2023) Optimizing reverse logistics for circular economy: strategies for efficient material recovery and resource circularity. Int J Soc Sci Except Res. https://doi.org/10.54660/ijsser.2023.2.1.16-31

De Angelis R (2024) Circular economy business models as progressive business models: evidence from circular start-ups. Bus Strateg Environ 33:6303–6314. https://doi.org/10.1002/bse.3821

Ezekwu E (2025) Sustainable supply chain practices: driving efficiency, reducing waste, and promoting circular economy models. Int J Sci Res Arch. https://doi.org/10.30574/ijsra.2025.14.1.0206

Farazi MZR (2024) Designing circular supply chains with digital technologies for competitive sustainability: an operation management perspective. Int J Sci Res Arch 13(1):2346–2359

Fraccascia L, Giannoccaro I, Agarwal A, Hansen EG (2019) Business models for the circular economy: opportunities and challenges. Bus Strateg Environ 28:430–432

Gabbur A (2020) Investigating the circular economy and its impact in the UK manufacturing sector. Aston Research Explorer

GIZ (2023) The circular economy ecosystem in the MENA region: assessment report

van de Haar A (2022) Supply chain traceability systems—robust approaches for the digital age. https://doi.org/10.1016/b978-0-323-91614-1.00010-1

Hammarlund C, Svensson K, Asche F, Bronnmann J, Osmundsen TC, Nielsen R (2024) Eco-certification in aquaculture—economic incentives and effects. Rev Fish Sci Aquaculture. https://doi.org/10.1080/23308249.2024.2440712

Hernandez Marquina MV, Le Dain MA, Joly I, Zwolinski P (2024) Exploring determinants of collaboration in circular supply chains: a social exchange theory perspective. Sustain Prod Consumption 50:1–19. https://doi.org/10.1016/j.spc.2024.07.017

Hidalgo-Crespo J, Riel A, Duberg JV, Bunodière A, Golinska-Dawson P (2024) An exploratory study for product-as-a-service (PaaS) offers development for electrical and electronic equipment. Procedia CIRP 122:521–526. https://doi.org/10.1016/j.procir.2024.01.076

Hobbs AJ (2021) Upskilling and retraining the adult workforce. https://doi.org/10.58248/hs28

Ingaldi M, Ulewicz R (2024) The business model of a circular economy in the innovation and improvement of metal processing. Sustainability. https://doi.org/10.3390/su16135513

Jadallah N, Bhatti F (2020) Political instability and sustainable green supply chain management. Manage Sci Lett 10(3):703–712. https://doi.org/10.5267/j.msl.2019.12.001

Jum'a L, Ikram M, Jabbour CJC (2024) Towards circular economy: a IoT enabled framework for circular supply chain integration. Comput Ind Eng 192:110194

Kanwal P, Quttainah MA, Kumar A (2024) Social impact of circular economy startups. Advances in business strategy and competitive advantage book series

Kesavaraj GKKP (2024) A study on enhancing sustainability in circular supply chains through green innovations. https://doi.org/10.52783/eel.v14i1.1179

Khompatraporn C (2021) Circular supply chain management. https://doi.org/10.1007/978-981-15-8510-4_21

Koval V, Arsawan IWE, Suryantini NPS, Kovbasenko S, Fisunenko N, Aloshyna T (2022) Circular economy and sustainability-oriented innovation: conceptual framework and energy future avenue. Energies 16(1):243

Mahjoob A, Alfadhli Y, Omachonu V (2023) Healthcare waste and sustainability: implications for a circular economy. Sustainability 15(10):7788

Marsala S, Melkionan K, Colvine A, Al Adwan R, Palaonda D, Haridi AM, Abdelmoula M (2022) CLUSTER (C_A.3.1_0014) WP3 (O.3.2)

Ministry of Local Government, P (2023) MoLG-JICA project for capacity development for solid waste management in Palestine: interim summary report on waste reduction/minimization planning

Moktadir MA, Kumar A, Ali SM, Paul SK, Sultana RS, Rezaei J (2024) Critical success factors for a circular economy: implications for business strategy and the environment. [Journal Name Pending]

Munir R, Fausiah R (2025) The role of circular economy in green entrepreneurship: sustainable business models. J Indonesian Scholars Soc Res. https://doi.org/10.59065/jissr.v5i1.173

Nika C-E, Expósito A, Kisser J, Bertino G, Oral HV, Dehghanian K, Vasilaki V, Iacovidou E, Fatone F, Atanasova N, Katsou E (2021) Validating circular performance indicators: the interface between circular economy and stakeholders. Water. https://doi.org/10.3390/W13162198

Nousala S, Metcalf G (2024) Industry 4.0 to industry 5.0: explorations in the transition from a techno-economic to a socio-technical future. Springer Nature, p 184

Onukwulu EC, Dienagha IN, Digitemie WN, Egbumokei PI, Oladipo OT (2025) Enhancing sustainability through stakeholder engagement: strategies for effective circular economy practices. South Asian J Soc Stud Econ 22(1):135–150

Onukwulu EC, Agho MO, Eyo-Udo NL (2022) Circular economy models for sustainable resource management in energy supply chains. World J Adv Sci Technol. https://doi.org/10.53346/wjast.2022.2.2.0048

Porwal PD (2023) Green business and environmental sustainability. Trends Bank, Account Bus 2(1):41–50

Rajčić V, Lin Y-H, Laban M, Tsikaloudaki K, Ungureanu V (2024) Cultural and societal challenges for circular strategies implementation. Sustainability. https://doi.org/10.3390/su17010220

Razia B, Awwad B (2022) A comprehensive review of blockchain technology and its related aspects in higher education. Technologies, artificial intelligence and the future of learning post-COVID-19: the crucial role of international accreditation, pp 553–571. https://doi.org/10.1007/978-3-030-93921-2_29

Razia AS, Alkababji MW, Ruzieh AS, Awwad BS, Razia BS (2024) Competitive advantage in light of AI to achieve innovative as a key of the SDGS: Palestinian insurance industry. J Lifestyle SDGs Rev 4(1):e01870–e01870. https://sdgsreview.org/LifestyleJournal/article/view/1870

Rejeb A, Rejeb K, Keogh JG, Süle E (2025) When industry 5.0 meets the circular economy: a systematic literature review. Circular Econ Sustain 1–32

Ruzieh AS, Awwad BS, Razia BS (2025) Review of logistics challenges within the construction industry. EuroMid J Bus Tech-Innovation 1–28. https://doi.org/10.51325/ejbti.v4i1.206

Ryan NA, Vladimirova D (2023) Introducing a novel framework for regenerative business. https://doi.org/10.26481/mup.2302.40

Shahvaroughi M, Farahani F, Ghasemi G (2024) How artificial intelligence plays a role in achieving sustainable development goals? Sustain Econ 2(3):66

Singh RJ (2022) Circular economy in supply chain. J Prod, Oper Manage Econ. https://doi.org/10.55529/jpome.23.35.45

Solomon NO, Simpa P, Adenekan OA, Obasi SC (2024) Circular economy principles and their integration into global supply chain strategies. Financ Account Res J 6(5):747–762

Surahman FD, Haryadi D (2021) The influence of maintenance and quality control on quality product. https://doi.org/10.36555/ALMANA.V5I1.1586

Taddei E, Sassanelli C, Rosa P, Terzi S (2024) Circular supply chains theoretical gaps and practical barriers: a model to support approaching firms in the era of industry 4.0. Comput Ind Eng 190:110049

van Erp LME (2022) Regenerative economy. https://doi.org/10.4324/9781003244196-25

Wen X, Siqin T (2020) How do product quality uncertainties affect the sharing economy platforms with risk considerations? A mean-variance analysis. Int J Prod Econ. https://doi.org/10.1016/J.IJPE.2019.107544

The Role of Policy and Regulation in Driving Sustainable Innovation

9

Asieh AkhlaghiMofrad

Abstract

This chapter discusses the indispensable role played by entrepreneurial policy and global regulation in promoting eco-innovation in companies, which build legal, economic, and social environments for sustainable actions. Referencing a variety of national, regional, and sectoral policies, including the types and levels of carbon pricing that have emerged from the EU ETS and from the Swedish carbon tax, it demonstrates how policies, not least different types of carbon pricing, have influenced corporate decision-making in respect of climate-friendly technology investment. The chapter also offers instruments for reporting the firms' sustainability actions and regulations that foster transparency and future planning. Some strategies, such as subsidies, green purchasing, and tax credits, are suggested to accelerate sustainable entrepreneurship opportunities and reduce risks. Also, the role of some international agreements, such as the Paris Agreement and UN Sustainable Development Goals, for global integration of sustainability purposes is highlighted. Finally, the chapter suggests some incentive and punitive strategies which could respectively promote innovation and prevention of actions and paths toward firms' social responsibility and eventually sustainability.

Keywords

Sustainability • Green innovation • International regulations • Carbon pricing

A. AkhlaghiMofrad (✉)
Girne American University, Kyrenia, Northern Cyprus
e-mail: Asiehmofrad@GAU.edu.tr

B. Awwad and R. A. Awwad (eds.), *Technology and Innovation Towards a Future of Business Sustainability*, Contributions to Environmental Sciences & Innovative Business Technology, https://doi.org/10.1007/978-981-95-6469-9_9

9.1 Introduction

This chapter explores the role of entrepreneurial policy at national, regional, and sectoral levels, and international regulations in promoting eco-innovation in firms through the construction of legal, economic, and social frameworks that encourage environmentally sound practices (Awwad 2024a). At a time when the climate is changing, the environment is degrading, and social demands are escalating, legislative and regulatory activity is increasingly becoming a critical tool to ensure business practices are consistent with sustainability goals (Awwad and Hamdan 2023).

The treatment begins with a review of some major policy strategies, particularly carbon-pricing measures, such as carbon taxes and ETS indicators. Such mechanisms include the environmental damage caused by releasing carbon into the atmosphere by firms and encouraging them to reduce their carbon activities. In some instances, such as Sweden's carbon tax and the European Union's Emission Trading System, we can observe how such strategies can impact firms' behavior, trigger their investment in clean technology, and help to notice other environmental issues. The chapter also reviews some environmental regulations with regard to setting limitations on emissions, usage of resources, and product safety, including mandatory ESG reporting rules, EU REACH, and the Corporate Sustainability Reporting Directive. The purpose of such measures is to promote clarity, conformity, and provision of any potential market opportunity that would lead to further investments and motivation of long-term planning (Nipper et al. 2022, 2025).

The focus is drawn on economic instruments such as subsidies, green public procurement, and tax credits among the instruments that would support and reduce the risks of failure in adopting sustainable and green business practices and green entrepreneurship. While non-compliant are being punished at the same time, their message is going out on accountability. International treaties like the Paris Agreement and the United Nations' Sustainable Development Goals (SDGs) are also important, because they globalize regulation and encourage both countries and companies to work on goals that support global sustainability objectives. The chapter works through strategic options, like the design and implementation of carbon border adjustments, technology-forcing rules, and public–private pacts on R&D. It's not a large lesson about multilateralism; it's about the coherence of policies, international cooperation, and inclusive systems of support for SME. Furthermore, the chapter maintains that effective regulation should not inhibit but rather enable business and recommend positive space for new business opportunities by providing an environment that is predictable, fair, and an opportunity creator. Policy and regulation are increasingly taking a lead role in the path to sustainability that has a strategic direction and a market-driven motivation to achieve.

Government policy and international regulations serve to constrain and enable corporate sustainability innovation by establishing expectations, lowering risks, and creating new opportunities. Government regulations play a vital role in ensuring the effective promotion and development of Corporate Social Responsibility

(CSR) and sustainable development by setting standards and policy structure to motivate business firms to operate in a way that will cover mutual benefits for society and the environment itself. With global challenges such as climate change, resource depletion, and social inequality becoming ever more pressing, the relevance of regulatory interventions by the state is increasing, orienting corporate conduct toward greater responsibility and sustainability in the long run (Collison 2024; Awwad et al. 2023).

At the international level, global legislation and agreements are important for the international promotion of Corporate Social Responsibility (CSR). There are some international treaties, such as the Paris Accords and UN's sustainable development goals, that propose widely recognized guidelines for sustainability. The aim of the guidelines is to impact the individual state regulations and motivate firms to take actions in line with global sustainability (Collison 2024). Establishing Legal Criteria for punishing damaging activities, e.g., through imposing restrictions on emissions, resource use, waste disposal, and environmental impact. A classic example that makes the point very well is the application of carbon-pricing mechanisms, like cap-and-trade systems or carbon taxes, that encourage businesses to reduce their greenhouse gas content (Aldy and Stavins 2012). Besides the legal standards and regulations, other strategies such as economic incentives, encouraging green investment and innovation, would be other tools that governments and international institutions consider for mitigating harmful environmental pollution.

Finally, the chapter introduces in detail some government policies and international regulations that determine the contexts and infrastructures for the corporate behavior from a legal, economic, and social dimension for sustainable business (Awwad 2024b). And then, potential strategies for driving sustainable innovation occur when they are supported by strong governance for innovative sustainability and stakeholder engagements, because these strategies might help to mitigate risks, reward leadership, and make sustainability a part of the economic fabric.

9.2 Government Policies and International Regulations

For better control of the resource consumption, waste, and emission, it is necessary to set some restrictive environmental regulations to direct firm's actions in line with sustainability goals (Kagan et al. 2003; Li et al. 2023). Generally, there are two types of action taken by the governments for implementing their objectives regarding environmental protection: one is to use a cap-and-trade strategy or impose a tax on pollution (financial incentive), the other is to impose more strict regulations on environmental control, such as the Clean Air, Water Acts, or Industrial Emissions Directive of the European Union. These strategies have known to be effective in the reduction of industrial pollutants by far (McManus 2009; Singhal 2018; Stavins 2003).

In addition to these policies, market-based instruments, in particular carbon pricing (i.e., in the forms of carbon taxes and cap-and-trade), have become increasingly popular. Economists say these mechanisms internalize the social costs of greenhouse gas emissions by making them a financial consequence for businesses (Razia et al. 2024). For example, recent meta-studies show that higher carbon prices reliably drive down CO_2 emissions, particularly when tied up with clear, long-term policy signals. Therefore, firms have the option of either changing their production procedures or choosing a low-carbon technology that promotes both their own interests and the environmental benefits (Kohlscheen et al. 2024; Döbbeling-Hildebrandt et al. 2024).

Next, the chapter discovers the impact of international regulations that could potentially shape and form firms' sustainable behavior and actions on legal, social, and economic grounds.

9.2.1 Product Standards and Sustainable Innovation

While legal frameworks address emissions, they also include product standards, such as eco-labeling, energy requirements, and limits of hazardous substances. For example, under the EU's REACH regulation, chemicals could be hazardous, and there is a requirement for a safety assessment for all chemicals, and proposing a ban on the substances that are proven hazardous unless explicitly authorized. Environmental economists call such regulations "technology-forcing" standards—they force companies to develop solutions that achieve minimal performance targets or lose the marketplace (Gerard and Lave 2007; Flammer 2021). As (Wang et al. 2023) suggest, in China, stringent environmental regulations tend to promote industrial low-carbon transitions in places with moderate technology and finance strength.

9.2.2 Mandatory Reporting and Corporate Transparency

Compelled disclosure requirements have developed as a further cornerstone in ESG governance. The EU's Corporate Sustainability Reporting Directive (CSRD), which comes into effect from 2024, is a milestone in mandatory, standardized ESG disclosure. CSRD forces thousands of large EU companies (and some non-EU ones) to publish audited data according to European Sustainability Reporting Standards (ESRS)—emissions, resource use, social impact, governance, etc. (Pantazi 2024). Reference (Lambert and Bleakley 2025) report that the tendency toward mandatory ESG reporting not only enhances data transparency but can act as an antidote to greenwashing, forcing more structured, fact-based disclosures.

The CSRD enforcement measures, such as publication bans, fines, and shareholder lawsuits, add strength to sustainability reporting and can help enforce compliance with it (Pantazi 2024). Experimental research confirms that mandatory ESG disclosures result in greater ESG performance and that investors have more trust in mandatory ESG disclosures, and hence investors have more trust in them (Krüger and Tang 2025; Awwad 2022). The legal consequences of CSRD, such as publication bans, fines, and shareholder litigation, help ensure compliance and give some substance to sustainability reporting. It is found in the empirical literature that companies subject to these mandatory ESG reporting requirements exhibit stronger ESG performance over time and investor confidence (Papafloratos and Pantazi 2025). By establishing emission thresholds, durability standards for eco-labeled products, and reporting obligations, governments set up a framework that limits actions and reconstructs behaviors toward sustainable innovation. As such, these regulatory and disclosure efforts allow companies to quickly identify strategic opportunities to advance and/or improve climate technologies, or more environmentally sustainable processes, or ESG-friendly products, and to outcompete rivals within the regulatory matrix.

9.2.3 Economic Incentives and Penalties

Successful policies don't suffice with simply requiring a regulation; they integrate both inescapable incentives and penalties so that rational economic choices achieve the same desired result without recourse to the question of "is global warming caused by man?". Such mechanisms can reduce the relative cost of adopting clean technologies or raise the cost of failing to comply—creating a potent nudge on firm behavior (Li et al. 2023).

The cost of investment in green technology may be reduced by incentives and subsidies from governments that take corrective taxes to offset the large negative externalities of pollution. Evidence from China shows that the clean technology investment is effectively enhanced by the direct R&D and environmental subsidies, especially for non-state-owned companies, by the increment of their R&D ability (Zhang et al. 2022).

In addition, renewable energy grants and tax benefits, such as feed-in tariffs and production tax credits, have had a profound impact on global energy markets. As one example of this, the U.S. investment tax credit (ITC), such as a 30% offset of the cost of a solar system, has led to multiple periods of tremendous growth in the deployment of solar, which has spiked every time the ITC was extended. Such policies not only spur uptake of the technology but also drive downstream innovation in manufacturing techniques, technological performance, and cost structures over time (Hu et al. 2019).

9.2.4 Green Public Procurement Programs

Public sector demand is a powerful magnifier of sustainable innovation. Green public procurement (GPP) initiatives oblige government agencies to buy eco-friendly products and services—establishing an early market for emerging clean technologies. When the governments promise to buy at most the low-carbon materials or energy-efficient appliances, they drive market risk out of innovative products, and deliver the fundamental scale needed to make sustainable products work financially (OECD 2019).

9.2.5 Environmental and Investment Tax Credits

In several countries, like Spain, Environmental and Investment tax credits have been effective in stimulating the adoption of cleaner production technologies by firms. The study of Spanish industries showed that credits that are directed at superior cleaner production technologies have enacted more than in the case of pollution control methods, especially at small firms that have capital constraints to attempt to execute any business decisions. This result highlights the importance of properly designed fiscal instruments that align firms' incentives not only in terms of innovation, but also in terms of compliance, as opposed to merely compensating for cost (Biggi et al. 2310).

9.2.6 Penalties and Fines

When firms expect to face severe monetary penalties or production bans, they start to cultivate sustainable practices and show high levels of conformity with the regulations. Imposing penalties would eventually lead to restoration and innovation of technological advancement, and a high level of participation and compliance could be effective tools for firms' financial driving objectives (Nyborg and Telle 2006).

Economic incentives, including subsidies, tax breaks, and green public procurement, and disincentives, including fines and remediation costs, are essential supplements to legal standards. By lowering costs, raising returns, and strengthening accountability, sound fiscal instruments unleash finance for sustainable change. The optimal results are not attained through individual instruments, but through integrated policy mixes, which align short-term compliance with long-term innovation goals, building robust markets for green technology (Zhang and Fu 2022).

9.2.7 International Agreements and Standards

Sustainable innovation and change are increasingly grounded on international agreements and standards that supply global targets, incentivize policy coherence, and apply institutional pressure on firms and governments to converge on sustainability goals in common (Biermann et al. 2017; Awwad 2025a). A legally binding international treaty, the Paris Agreement (2015) has transformed corporate behavior to a considerable degree. Research suggests that it has enhanced worldwide climate awareness, affecting firm capital structure and disclosures. Studies show that the ratification of nations has been followed by an increase in litigation related to climate matters, as well as legal obligations requiring corporate directors to take into account climate risk. In addition, investor coalitions are now calling for Paris-aligned strategies, endorsing disclosure criteria such as TCFD and asking companies to set emissions targets in line with the Agreement. These vehicles all serve to increase the cost of doing nothing and drive systemic inclusion of climate within business strategy (Kölbel et al. 2020; Setzer and Vanhala 2019; Sullivan 2017). The UN SDGs, created in 2015, are the ultimate blueprint for sustainability. While not binding documents, they are becoming increasingly enshrined in standards of reporting and corporate governance. Research findings suggest that the motivations behind firms' SDG-related reporting are competitive and legitimacy-seeking, evidencing the institutional pressures they confront—especially those of sectors and countries with a strong governance demand. The high performing ones are showing stronger linkage to bottom line results when the SGDs are being integrated more fully compared with a "sporadic" approach. These frameworks set the norms and material decisions that lead businesses to work for systemic reform as guided by global priorities (Waal and Thijssens 2020; Pizzi et al. 2021; Rosati and Faria 2019). Another agenda in trade agreements and ESG clauses is the environmental, social, and governance clauses—that push member states to enforce environmental and labor standards across international supply chains. While these clauses are nascent ones, they represent a good movement toward world harmonization and prevent compliance fragmentation and race to the bottom. Through the incorporation of ESG into cross-border commerce, international trade law serves as an effective supplement to hard environmental treaties and domestic laws, broadening the regulatory reach to multinational producers and suppliers at an international level (Monteiro 2022; Morin and Jinnah 2018; Lejárraga 2014). Furthermore, an abundance of global standards built out common ESG measurements and corporate expectations. The Task Force on Climate-related Financial Disclosures (TCFD), supported by the Financial Stability Board, provides a globally accepted template for reporting on climate-related risk and will become mandatory in certain jurisdictions, including the UK. The other, the Science Based Targets initiative (SBTi), allows companies to set emissions targets in line with the Paris Agreement, and so far, more than 10,000 organizations have signed up. These institutional arrangements mitigate greenwashing, increase comparability, and standardize reporting across the international systems (Kölbel et al. 2020; Amel-Zadeh and Serafeim 2018; Boffo and Patalano 2020).

9.2.8 Institutional Pressures and Corporate Norms

Neo-institutional theory describes the normative and mimetic pressures of policymakers, investors, and social expectations, which together influence corporate actors to be concerned with sustainability. The literature has found empirical evidence that firms are pressured to include SDGs in order to maintain legitimacy and satisfy the needs of stakeholders. ESG and SDG alignment is also strategic alignment and contributes to superior financial performance through legitimacy, reputation, and operational efficiency gains realized as companies exploit this source of competitive advantage (Rosati and Faria 2019; Brammer et al. 2012; Fernando and Lawrence 2014). Trade-focused instruments such as environmental clauses in free trade agreements can act to coordinate a minimum level of sustainability expectations across national borders (Brandi et al. 2020). A cross-country analysis of APEC treaties revealed that enforceable environmental clauses helped alleviate the pollution haven theory and facilitated balancing economic integration and environmental preservation (Lagria and Quismorio 2022). This regulatory uniformity encourages companies to innovate in order to comply with higher international standards and to prevent unnecessary regulatory burdens (Monteiro 2022; George and Yamaguchi 2018; Baghdadi et al. 2013). An equitable market should also ensure that sustainability standards do not unduly penalize smaller enterprises. Empirical evidence from the EU indicates that support through finance, training, and streamlined requirements for SMEs gave an opportunity for all to use the adoption of sustainable practices for business in all sizes to be equal, allowing to remain competitive without stifling younger and smaller innovators. Nature and content of the business case made to internal stakeholders. Such accessibility frameworks are critical for bringing landscape-wide innovation to the doorsteps of the corporations (Baldo 2017; Jansson et al. 2017; Biondi et al. 2000). Beyond legal prescriptions, public pressure and stakeholder demands greatly influence corporate sustainability moves. Public policies that increase transparency make this scrutiny possible, allowing the investor, consumer, employee, and civil society to demand corporate behavior that is both responsible and accountable. (Aguilera et al. 2007; Delmas and Toffel 2008; Fung et al. 2007). Transparency requirements—including requirements for environmental reporting, stakeholder consultations, and public reporting portals—can be held up as drivers for societal involvement. When companies are forced to disclose their emission records or usage of resources, it would allow NGOS, media, and community organizations to campaign against the underperforming ones. For example, studies show that required pollution disclosures in India and China reduced smog and improved firm performance by increasing local public awareness and participation (Lipscomb and Mobarak 2017; He and Sun 2021; Pargal and Wheeler 1996). Moreover, investors, especially institutional investors, increasingly use their capital to drive the firm toward sustainability. Climate risk and ESG strategy-focused resolutions and sustainability-oriented proxy voting have become viable methods over the last decade. According to large-scale empirical analyses, climate proposals raised by investors lead to a significant increase in the probability that firms will accept and implement GHG goals and sustainability plans (Flammer 2021; Grewal et al. 2016; Dyck et al. 2019; Krueger et al. 2020). Additionally, consumers' desire for ecologically friendly and morally produced products strengthens the pressure in the market. Empirical research supports the thesis that eco-branding and environmental labeling influence purchasing habits in a beneficial manner. Due to these findings, firms step up their game in product creation, packaging, and supply chain changes. Most prominently, using "Fair Trade," "Energy Star," and "USDA Organic" labels is a result of market-driven regulations (Pargal and Wheeler 1996).

9.3 The Potential Strategies for Driving Sustainable Innovation

9.3.1 Carbon Pricing Strategies

Carbon pricing has grown to be an important policy instrument to drive down greenhouse gas (GHG) emissions and to spur low-carbon innovation. The most effective instruments include carbon taxes, emissions trading schemes (ETS), and carbon border adjustment mechanisms (CBAM) aimed at internalizing the social cost of carbon and triggering economic incentives for reducing emissions. Carbon taxes work by charging a price on each ton of CO_2 emitted, prompting polluters to cut emissions when and where it is least expensive to do so. Studies indicate that the establishment of a predictable and steadily rising carbon tax schedule is a necessity in order to provide a stable investment climate to support clean energy and innovation (Metcalf and Stock 2020). In addition, allocating carbon tax receipts to the research and development (R&D) of clean energy, low-carbon infrastructure, or compensatory rebates to poorer sections of the population improves political acceptance and mitigates distributional fairness (OECD 2019). Policy measures such as a carbon tax, which has been successfully implemented in countries like Sweden and Canada, raise funds for energy transition and compensate lower-income households (Carl and Fedor 2016).

Another way to allow more flexibility is provided by cap-and-trade systems, or emissions trading schemes, in which the total amount of emissions is capped, and the price of emissions permits is determined by the market. These instruments guarantee environmental results by imposing absolute emission constraints, and they also introduce efficiency in terms of cost through the character of tradable permits. If properly designed—with hard and declining caps, and transparent monitoring—ETS systems are capable of driving large cuts in emissions (Ellerman et al. 2010). The first and the biggest system at the European Union level is the EU ETS, and the evidence suggests that it indeed reduced emissions in some of its main industrial sectors (Martin et al. 2016) and acts as an example for other regulatory jurisdictions. To increase the impact of measures, some areas have started trading in carbon credits across borders (not only enlarging potential market size but also harmonizing global efforts and overcoming concerns about competitiveness (Flachsland et al. 2020).

Finally, Carbon Border Adjustment Mechanisms (CBAM) are a relatively new, but increasingly prominent, tool to address carbon leakage—the shifting of polluting industries to jurisdictions with weaker environmental standards. CBAMs apply taxes to the imported goods proportional to their CO_2 content in order to prevent domestic producers that are subject to low emission standards from being undercut by foreign competitors (Cosbey et al. 2019). By leveling the playing field, they support domestic carbon pricing and can signal to other trading partners that they could benefit from having similar climate policies. The European Union's upcoming CBAM is a leading case that could become a global precedent (European Commission 2021). The carbon price measures provide a strong set of tools to get economic activity in line with climate ambition. If deployed in thoughtful concert with one another, carbon taxes, cap-and-trade systems, and border adjustments can spur systemic decarbonization, redound to innovation and fairness, and encourage collaboration among countries.

9.3.2 Environmental Regulation Strategies

Robust environmental regulation is critical to achieve systemic sustainability transitions in sectors, economies, and international supply chains (Awwad 2025b). Governments have combined a variety of regulatory instruments in devising responses to climate change and environmental harm, such as performance standards, mandatory environmental reporting, and the banning or phasing out of harmful activities. Taken together, these measures lift the environmental baseline, encourage innovation, and act as drivers for market expectations.

Performance standards, in the form of energy efficiency and emission standards, for industrial sectors, transportation systems, and household appliances, are among the most common regulatory instruments. These standards internalize the costs of environmental externalities by making producers and consumers responsible for meeting some minimal level of environmental performance. As Gillingham and Stock (2018) notes, energy efficiency standards on U.S. vehicles and buildings can cut greenhouse gas emissions at a low net cost over time. Beyond these static standards, governments can adopt technology-forcing regulations, which establish challenging future objectives that only innovation can meet. Regulations of this kind, for example, the E.U. CO_2 emission limits for new cars, force the industry to introduce and take up cleaner technologies (Requate 2005).

A third heavy regulatory club is mandatory environmental reporting. Making firms responsible for disclosing their carbon footprints on the environment and reporting their actions would enhance the clarity of their initiatives. There are some action plans, such as the European Corporate Sustainability Reporting Directives and the Task Force on Climate-related Financial Disclosures, that are proactive in implementing environmental risk reporting (Serafeim 2020). As Clark and Hebb (2016) advocate, compulsory ESG disclosures not only decrease information asymmetries but also encourage firms to manage environmental risks proactively. Another form of environmental control is the prohibition or the gradual elimination of destructive environmental practices. The examples are many: a planned ban of Internal Combustion Engine Vehicles (ICEVs) in various countries by 2035 to speed up the move to electric mobility. Also, phasing out the fossil fuel subsidies, which reached a global level of more than $ 450 billion in 2022, is fundamental to align the public finance with the environmental goals (International Energy Agency (IEA) 2023). Withdrawing subsidy support for high-emitting industries can create an even playing field for green technologies, and it sends a strong signal about the direction of future energy systems. As Koplow and Steenblik (2008) suggests, subsidy reform results in lower emissions and also an improvement in economic efficiency by correcting price distortions.

In short, the strategies of environmental regulations are of high importance in determining sustainable economic conduct. Performance standards, transparency requirements, and bans or subsidy reform correct the systemic incentives. Used together, those tools create a broad-based policy framework to direct countries and companies to a future more resilient to environmental change and more dependent on low-carbon innovation.

9.3.3 Incentive-Based Strategies

An incentive-based strategy would be subsidies, tax reduction, and public ownership policies offered by governments. These incentives can lower the monetary risks involved with innovation and give way to more favorable market conditions (Marsal et al. 2024). An integration of these polices would encourage the private sector to further invest in green technology, especially when there is financial instability in the market and during the earlier stages of sustainable development.

Besides focusing purely on research, government funding on circular economy innovations, carbon capture projects, and clean hydrogen production (with minimal greenhouse gas emission) should be on the agenda (Mazzucato 2016; Khoffash and Awwad 2024). Furthermore, initiatives like promoting public–private partnerships that are willing to test new technologies at scale could help with popularization and knowledge management in the field (OECD 2020). The process of public authorities producing and purchasing goods and services with lower environmental impacts (Green) is called Green public procurement, which is a high-impact strategy for the reduction of public services' environmental footprint and the authorization and adoption of green practices by the private sector in the market (Testa et al. 2016). Sustainable investments are cheaper with tax incentives and credits. The attractiveness and behavior-shaping power of smart infrastructure come across through tax breaks for renewable energy deployment, circular economy projects, or sustainable building retrofits over time (Carley and Konisky 2020).

Ultimately, these incentive-based models reduce innovation barriers, remove risk to green investment, and contribute to an ecosystem that supports sustainable economic transformation.

9.4 Conclusion

In this chapter, we explore the impact of national, regional, and international policies on eco-innovation, which define the legal, economic, and social environments that lead companies toward an environmentally friendly management. Given climate change, ecological decline, and the need for a healthier society, regulation has become a key tool through which to align commercial behavior with what we want to achieve.

One of the key policy options is the implementation of carbon-pricing tools such as a carbon tax or Emissions Trading System (ETS), which can help to internalize the environmental cost of carbon emissions and incentivize corporations to act in a more environmentally friendly manner. Case studies like the EU ETS and the Swedish carbon tax have demonstrated that they can provide incentives for investments in cleaner technology.

The chapter also discusses environmental laws that put legal limits on emissions, resource consumption, and product safety, such as EU's REACH regulation and the Corporate Sustainability Reporting Directive (CSRD). These are the frameworks of transparency, investor confidence, and long-term strategic planning. Economic tools, including subsidies, green public procurement, and tax incentives, are noted as instruments to reduce economic risks and promote sustainable entrepreneurship. Additionally, penalties for violations would reinforce responsibility.

Global climate policies and international agreements, such as the Paris Agreement and the UN SDGs, are shown to induce the globalization of sustainability projects, harmonization of norms, and domestic policy changes. Other Green Deal policies analyzed in the report include carbon border adjustments, technology-forcing legislation, and public–private R&D partnerships. Whereas those strategies are based on integrated policies, international cooperation, and fan-out support systems—notably for small and medium-sized enterprises (SMEs).

Finally, the chapter concludes that regulation should not stifle progress but enable it by providing an environment that is transparent, consistent, and full of opportunity. Standards and decreasing the degree of uncertainty make policy and regulation a key enabler for corporate sustainability, Corporate Social Responsibility (CSR), and global sustainable development.

References

Aguilera RV, Rupp DE, Williams CA, Ganapathi J (2007) Putting the S back in corporate social responsibility: a multilevel theory of social change in organizations. Acad Manag Rev 32(3):836–863

Aldy JE, Stavins RN (2012) The promise and problems of pricing carbon: theory and experience. J Environ Dev 21(2):152–180

Amel-Zadeh A, Serafeim G (2018) Why and how investors use ESG information: evidence from a global survey. Financ Anal J 74(3):87–103

Awwad BS (2022) The role of environmental disclosure in enhancing the market value of industrial and service companies listed on the Palestinian stock exchange. Int J Entrepr 26(2):1–12

Awwad BSAL (2024a) Governance with relationship between entrepreneurship and economic growth in Palestine. Int J Law Manag 66(2):259–287

Awwad B (2025b) Climate finance and sustainability: a story told through the Journal of Cleaner Production. EuroMid J Bus Tech-Innov 4:31–47

Awwad BSAL, Hamdan AMM (2023) Saudi Arabia's post-oil economic prospects: economic diversification, entrepreneurship and women's empowerment. J Hum Resour Sustain Stud 11:903–941

Awwad BS, Alkababji MW, Zidan S (2023) Governance and social responsibility: Palestinian investment fund. Bus Econ Res 13(4):185–215

Awwad B (ed) (2024) The AI revolution: driving business innovation and research. Springer

Awwad BS (ed) (2025) Sustainability in light of governance and artificial intelligence applications. Emerald Group Publishing
Baghdadi L, Martinez-Zarzoso I, Zitouna H (2013) Are RTA environmental provisions beneficial? Evidence from NAFTA. Rev Int Econ 21(4):847–861
Biermann F, Kanie N, Kim RE (2017) Global governance by goal-setting: the novel approach of the UN Sustainable Development Goals. Curr Opin Environ Sustain 26–27:26–31
Biggi G, Mina A, Tamagni F (2023) There are different shades of green: heterogeneous environmental innovations and their effects on firm performance. arXiv preprint arXiv:2310.08353
Biondi V, Frey M, Iraldo F (2000) Environmental management systems and SMEs: motivations, opportunities and barriers related to EMAS and ISO 14001 implementation. Greener Manag Int 29:55–69
Boffo, R., & Patalano, R. (2020). ESG investing: practices, progress and challenges. OECD.
Brammer S, Jackson G, Matten D (2012) Corporate social responsibility and institutional theory: new perspectives on private governance. Soc Econ Rev 10(1):3–28
Brandi C, Schwab J, Berger A, Morin JF (2020) Do environmental provisions in trade agreements make exports from developing countries greener? World Dev
Carl J, Fedor D (2016) Tracking global carbon revenues: a survey of carbon taxes and emissions trading systems. Clim Policy 16(6):709–731
Carley S, Konisky DM (2020) The justice and equity implications of the clean energy transition. Nat Energy 5(8):569–577
Clark GL, Hebb T (2016) Why investors need to understand ESG. Oxford University Press
Collison N (2024) The role of government regulation in promoting corporate social responsibility and sustainable development. J Int Bus Res 22(1):45–63
Cosbey A, Droege S, Fischer C, Munnings C (2019) Developing guidance for implementing border carbon adjustments: lessons, cautions, and research needs. Rev Environ Econ Policy 13(1):3–22
Del Baldo M (2017) Sustainable corporate responsibility for SMEs in the European Union: a framework for supporting responsible business practices. Eur J Sustain Dev 6(3):347–360
Delmas M, Toffel MW (2008) Organizational responses to environmental demands: opening the black box. Strateg Manag J 29(10):1027–1055
Döbbeling-Hildebrandt N, Miersch K, Khanna TM, Bachelet M, Bruns SB, Callaghan M, Minx JC et al (2024) Systematic review and meta-analysis of ex-post evaluations on the effectiveness of carbon pricing. Nat Commun 15(1):4147
Dyck A, Lins KV, Roth L, Wagner HF (2019) Do institutional investors drive corporate social responsibility? International evidence. J Financ Econ 131(3):693–714
Ellerman AD, Convery FJ, de Perthuis C (2010) Pricing carbon: the European Union emissions trading scheme. Cambridge University Press
European Commission (2021) Proposal for a regulation establishing a carbon border adjustment mechanism
Fernando S, Lawrence S (2014) A theoretical framework for CSR practices: integrating legitimacy theory, stakeholder theory and institutional theory. J Theor Account Res 10(1):149–178
Flachsland C, Pahle M, Burtraw D, Edenhofer O, Elkerbout M, Fischer C, Zetterberg L (2020) How to avoid history repeating itself: the case for an EU Emissions Trading System review. Energy Policy 142:111485
Flammer C (2021) Corporate green bonds. J Financ Econ 142(2):499–516
Fung A, Graham M, Weil D (2007) Full disclosure: the perils and promise of transparency. Cambridge University Press
George C, Yamaguchi S (2018) Assessing the effectiveness of environmental provisions in regional trade agreements: an analytical toolkit. OECD Trade and Environment Working Papers, No. 2018/01
Gerard D, Lave L (2007) Experiments in technology forcing: comparing the regulatory processes of US automobile safety and emissions regulations. Int J Technol Policy Manag 7(1):1–14
Gillingham K, Stock JH (2018) The cost of reducing greenhouse gas emissions. J Econ Perspect 32(4):53–72. https://doi.org/10.1257/jep.32.4.53

Grewal J, Serafeim G, Yoon A (2016) Shareholder activism on sustainability issues. Harvard Business School Accounting & Management Unit Working Paper No. 16-003
He G, Sun C (2021) Can transparency policies reduce pollution? Evidence from China's mandate to disclose air pollution information. J Environ Econ Manag 108:102425
Hu Z, Lin B, Zhang F (2019) Effect of government subsidies on renewable energy investments: the threshold effect. Energy Policy 132:156–166
International Energy Agency (IEA) (2023) World energy outlook 2023. https://www.iea.org/reports/world-energy-outlook-2023
Jansson J, Nilsson J, Modig F, Hed Vall G (2017) Commitment to sustainability in small and medium-sized enterprises: the influence of strategic orientations and management values. Bus Strateg Environ 26(1):69–83
Kagan RA, Gunningham N, Thornton D (2003) Shades of green: business, regulation, and environment. Stanford University Press
Khoffash S, Subhi Awwad B (2024) The challenges and opportunities of green financing in palestine from the perspective of banks. J Infrastruct Policy Dev 8(2):1–20. https://doi.org/10.24294/jipd.v8i12.8800
Kohlscheen E, Moessner R, Takats E (2024) Effects of carbon pricing and other climate policies on CO_2 emissions. arXiv preprint arXiv:2402.03800
Kölbel JF, Heeb F, Paetzold F, Busch T (2020) Can sustainable investing save the world? Reviewing the mechanisms of investor impact. Organ Environ 33(4):554–574
Koplow D, Steenblik R (2008) Subsidies to fossil fuel production: an inventory of policies and estimates. OECD Global Subsidies Initiative
Krueger P, Sautner Z, Starks LT (2020) The importance of climate risks for institutional investors. Rev Financ Stud 33(3):1067–1111
Krüger, Sautner, Tang (2025) The effects of mandatory ESG disclosure around the world. J Account Res
Lagria RF, Quismorio BA (2022) Analyzing trends in APEC using data analytics. PIDS Discussion Paper Series
Lambert J, Bleakley W (2025) Does mandatory greenhouse gas emissions reporting program deter corporate greenwashing? J Environ Manag 373:123740
Lejárraga I (2014) Deep provisions in regional trade agreements: how multilateral-friendly? OECD Trade Policy Papers, No. 168
Li F, Liao X, Li Y (2023) The role of financial incentives in promoting green technology innovation: evidence from Chinese listed firms. Environ Econ Policy Stud 25:789–812
Lipscomb M, Mobarak AM (2017) Decentralization and pollution spillovers: evidence from the re-drawing of county boundaries in China. Rev Econ Stat 99(3):467–483
Marsal H, Hamdan A, Awwad BSALA, Mohamed M (2024) The impact of mentorship and funding support on stimulating entrepreneurship motivation among family members. Eur J Fam Bus 14(1):117–130
Martin R, Muûls M, de Preux LB, Wagner UJ (2016) Industry compensation under relocation risk: a firm-level analysis of the EU ETS. Am Econ Rev 106(5):387–391
Mazzucato M (2016) From market fixing to market-creating: a new framework for innovation policy. Ind Innov 23(2):140–156
McManus CG (2009) The effects of clay and a freshwater coagulant on Daphnia magna. Wetland Sci Pract 26(2):4–7
Metcalf GE, Stock JH (2020) The macroeconomic impact of Europe's carbon taxes. Brookings Pap Econ Activ 1–67. https://www.brookings.edu/wpcontent/uploads/2020/03/Metcalf_Stock_conference_draft.pdf
Monteiro J-A (2022) Environmental provisions in regional trade agreements: promises and Pitfalls. World Trade Rev 21(4):501–521
Morin JF, Jinnah S (2018) The North-South politics of global sustainability standards. Glob Environ Polit 18(3):47–66
Nipper M, Ostermaier A, Theis J (2022) Mandatory disclosure of standardized sustainability metrics: the case of the EU taxonomy regulation. Sustainability 17(12):5336

Nipper M, Ostermaier A, Theis J (2025) Navigating CSRD reporting: turning compliance into sustainable development with science-based metrics. Environ Dev 54:101138
Nyborg K, Telle K (2006) Firms' compliance to environmental regulation: is there really a paradox? Environ Resource Econ 35(1):1–18
OECD (2019) Taxing energy use 2019: using taxes for climate action. OECD Publishing
OECD (2020) The digitalisation of science, technology and innovation: key developments and policies
Pantazi T (2024) The introduction of mandatory corporate sustainability reporting in the EU and the question of enforcement. Eur Bus Organ Law Rev 25(3):509–532
Papafloratos T, Pantazi T (2025) A systematic review of the effects of mandatory corporate sustainability reporting. Sustainability 17(12):5336
Pargal S, Wheeler D (1996) Informal regulation of industrial pollution in developing countries: evidence from Indonesia. J Polit Econ 104(6):1314–1327
Pizzi S, Rosati F, Venturelli A (2021) The determinants of business contribution to the 2030 Agenda: introducing the SDG Reporting Score. Bus Strateg Environ 30(1):404–421
Razia AS, Alkababji MW, Ruzieh AS, Awwad BS, Razia BS (2024) Competitive advantage in light of AI to achieve innovative as a key of the SDGS: Palestinian Insurance industry. J Lifestyle SDGs Rev 4(1):e01870–e01870
Requate T (2005) Dynamic incentives by environmental policy instruments—a survey. Ecol Econ 54(2–3):175–195. https://doi.org/10.1016/j.ecolecon.2004.12.033
Rosati F, Faria LGD (2019) Business contribution to the Sustainable Development Agenda: organizational factors related to early adoption of SDG reporting. Corp Soc Responsib Environ Manag 26(3):588–597
Serafeim G (2020) How to build enduring commitment to long-term goals. Harv Bus Rev 98(1):122–131
Setzer J, Vanhala LC (2019) Climate litigation: a new class of environmental governance? J Environ Law 31(3), 309–338. Provides evidence that Paris Agreement ratification has contributed to increased climate-related litigation and legal duties
Singhal P (2018) Environmental regulations: Lessons from the command-and-control approach (No. 124). DIW Roundup: Politik im Fokus
Stavins RN (2003) Market-based environmental policies: what can we learn from US experience (and related research)?
Sullivan R (2017) Fiduciary duty in the 21st century. PRI, UNEP FI & The Generation Foundation. Links the Paris Agreement to the evolution of fiduciary duties regarding climate risk
Testa F, Annunziata E, Iraldo F, Frey M (2016) Drawbacks and opportunities of green public procurement: an effective tool for sustainable production. J Clean Prod 112:1893–1900
Van der Waal JW, Thijssens T (2020) Corporate involvement in sustainable development goals: exploring the territory. J Clean Prod 252:119625
Wang, Chen, Dong, Zhang, Shi (2023) Find these effects are most pronounced in regions with adequate technological and financial capacity
Zhang R, Fu Y (2022) Technological progress effects on energy efficiency from the perspective of technological innovation and technology introduction: an empirical study of Guangdong, China. Energy Rep 8:425–437
Zhang W, Liu X, Wang D, Zhou J (2022) Digital economy and carbon emission performance: evidence at China's city level. Energy Policy 165:112927

How Green Finance Innovation Drives Sustainability in North African Countries

10

Ala Fathi Assi and Lydia Iradukunda

Abstract

This study examines the impact of natural resource availability, renewable energy, green technology, and green finance on the financial development of North African economies, specifically Egypt, Algeria, Morocco, and Tunisia, from 1999 to 2021. The research uses advanced panel data econometric techniques and a Hausman test to identify the most appropriate model. Results show that green finance has a positive long-term relationship with financial development, underscoring the importance of environmentally focused financial instruments and policies in fostering sustainable developments. However, its short-run effect is not statistically significant, indicating a time lag in translating green finance into tangible financial sector improvements. Green technology has a robust positive influence on financial development across both short- and long-run horizons, highlighting the immediate and sustained benefits of technological innovation in driving financial development. Renewable energy contributes positively to financial development, with significance emerging only in the long run, suggesting that investments in clean energy infrastructure require longer periods to impact financial systems substantially. Natural resource availability shows a significant positive effect only in the short run, reflecting the initial economic stimulus provided by resource exploitation but cautioning against over-reliance on natural resources for long-term financial development. The paper offers pragmatic policy recommendations, including prioritizing strategic investments in green finance mechanisms and technological innovation to catalyze sustainable financial development.

A. F. Assi (✉) · L. Iradukunda
Near East University, TRNC, Mersin, Turkey
e-mail: alafathi.assi@neu.edu.tr

L. Iradukunda
e-mail: 20234712@std.neu.edu.tr

B. Awwad and R. A. Awwad (eds.), *Technology and Innovation Towards a Future of Business Sustainability*, Contributions to Environmental Sciences & Innovative Business Technology, https://doi.org/10.1007/978-981-95-6469-9_10

Keywords

Financial development • Green technology • Natural resources • Sustainability • Green finance

10.1 Introduction

It is difficult to create sustainable economies based on abundant resources, particularly in places with little institutional and financial development (Awwad 2025a, b). The unresolved issues of poverty—over-reliance on resources, economic survival, and environmental exploitation—are faced by North African nations including Egypt, Algeria, Morocco, and Tunisia. These nations nevertheless face fundamental problems including poor economic diversification, slowed economic growth, and pollution while having an abundance of natural resources. In addition to systemic obstacles, there are a number of strategic concerns that need to be addressed in order to develop carbon-free economies.

Natural resources are crucial for global economic growth, especially in underdeveloped nations. However, the resource curse hypothesis highlights problems with financial systems and economic growth in resource-boomed nations (Khan et al. 2020a). Economists become overly focused on exporting and resource management, leading to corruption, bad governance, and unstable growth. Nations with abundant resources often become economically dependent on these resources, neglecting other productive sectors. Consistent financial growth is difficult to achieve due to hydrocarbons being the main export. The Dutch disease, caused by high exchange rates, reduces non-resource businesses' competitiveness. The "Resource Curse" requires robust institutional frameworks to ensure long-term development-enhancing investments while maintaining financial stability (Mlachila and Ouedraogo 2020).

Renewable energy can address environmental issues and resource country syndrome, contributing to sustainable development globally according to UNDP 2021. Morocco, with its Noor Ouarzazate Solar Complex, is leading the way in renewable energy use. However, appropriate policies and significant investment are needed for renewable energy to be successful (Awwad 2025a, b). Investments in renewable energy sources boost energy security and the financial industry, promoting innovation and entrepreneurship. The Noor Ouarzazate Solar Complex is a prime example of this.

Green technology promotes environmentally friendly inventions, such as low-carbon products, improved waste disposal, and energy-efficient machinery, which can lead to long-term economic growth. Countries with abundant natural resources, like North Africa, can address local demands in green technology markets, such as reducing desert expansion and increasing water shortages. However, without sufficient funding, the adoption and proliferation of green technologies can be challenging.

Green finance is essential for supporting environmentally friendly operations by allocating funds for environmental protection initiatives (Awwad 2022). It helps move the economy towards a more environmentally friendly future by funding energy efficiency, sustainable infrastructure, and renewable energy projects. However, North Africa's green financing markets are still young and have a narrow market structure with inconsistent regulations. Governments, financial institutions, and investors must work to improve investment conditions and reduce the risk of financial instability to boost the supply of green financing (Khoffash and Awwad 2024).

This study contributes significantly to the scholarly and policy debates surrounding the economic growth of nations with abundant natural resources, such as Egypt, Algeria, Morocco, and Tunisia. It presents a conceptual framework that integrates natural resources, renewable energy, green technology, and green finance to investigate their link with financial development. The research advances the body of information already in existence by examining the relationship between these factors and financial systems in nations with abundant natural resources.

This study adds to the body of knowledge by examining scholarly and political discussions surrounding the concepts of sustainability and development in North African nations with abundant natural resources. While concurrently advancing environmental sustainability, this study fills the knowledge vacuum and improves our understanding of how these issues affect financial systems in resource-rich nations. In order to offer recommendations tailored to the particular situation, the study also looks at the economic and environmental issues in order to improve financial growth in North Africa.

10.2 Theoretical Section

Most resource-rich developing countries suffer from the curse of these resources, rather than their blessing. These economies experience slow economic growth due to weak fiscal policies and severe environmental degradation due to neglect of environmental policy implementation. This leads to economic distortion, coupled with environmental distortion and severe weaknesses in their financial structures. Financial growth plays a central role in promoting green economic growth by enabling efficient capital allocation, enhancing access to credit, and encouraging risk management mechanisms. However, in resource-dependent economies, excessive reliance on resource rents often hinders financial sector development, discouraging private sector investment and financial sector reforms. The theoretical integration of the "resource curse" framework with sustainable financial development highlights the importance of transitioning from traditional resource dependence to a more sustainable economic model. This perspective shifts the narrative from one of inevitable financial backwardness to one that recognizes the potential for harnessing resource wealth to achieve long-term financial sector growth, provided the appropriate institutional and policy frameworks are in place.

10.3 Literature Review

10.3.1 Financial Development and Natural Resources

Interest in financial operations that support resource efficiency and environmental sustainability has increased. Studies have shown that natural resources have a long-term detrimental effect on financial sustainability, as supported by the "financial resource curse" theory. Oben (2022) investigated the influence of natural resources on global financial development from 1980 to 2019, finding evidence in favor of this theory. Sun et al. (2020) found evidence in favor of the resource curse theory. Yıldirim et al. (2020) found that rising oil revenues had a long-term beneficial impact on financial development in 16 developing nations. Shahbaz et al. (2024) used the ARDL regression method to examine the connection between foreign capital accumulation and natural resource earnings in resource-rich nations from 1970 to 2016. They found that financial growth has a beneficial short- and long-term impact on the link between profits from natural resources and foreign capital accumulation. Khan et al. (2020b) studied the relationship between financial development and natural resource income in 87 emerging and developing economies, revealing that the riches of natural resources impede the advancement of finance in rising and developing nations.

10.3.2 Financial Development and Renewable Energy

Research on the relationship between renewable energy and financial development is ongoing. Some studies have shown a negative association, while others have shown a favorable relationship. Dimnwobi et al. (2022) found that financial development positively affects Turkey's use of renewable energy. Rezagholizadeh and Abdi (2022) found that financial development significantly aids both industrialized and developing nations in their renewable energy technology advancements. Saygin and Iskenderoglu (2022) found that financial growth does not impact renewable energy usage in emerging nations. Skare et al. (2023) found that energy development and macroeconomic conditions are crucial for attracting capital and improving financing conditions for renewable energy businesses. Prempeh (2023) found that Ghana's financial development encouraged renewable energy use, while Chireshe (2021) confirmed a favorable correlation between 2000 and 2016 in some sub-Saharan African nations.

10.3.3 Financial Development and Green (Technology and Finance)

By decreasing risk, enhancing capital allocation, and combining investments, financial growth can promote economic advancement (Nawaz et al. 2021; Awwad 2024). Financial infrastructure can affect energy consumption and boost economic expansion. Increased financial development leads to increased venture capital, financial market expansion, and economic development (Abidin et al. 2021; Hwang et al. 2022). It promotes the use of homes, vehicles, and electrical devices (Zhang et al. 2022). These financial sector services raise carbon emissions because of rising society consumption and declining environmental quality. Shahbaz et al. (2022) contend that financial markets are connected to investment activity and economic development through two methods. First, they entail using well-established financial channels to raise cash for high-yield companies. The second is that they increase output. They illustrate how financial development boosts asset diversity, liquidity, and the flow of capital to more lucrative businesses. There are three ways that financial development influences energy consumption: the wealth effect, the business effect, and the direct effect. A direct influence is linked to an environment with strong financial growth. Customers may find it easier and more affordable to borrow money to buy durable products and use more energy (Khan et al. 2022). There is a trade effect when financial development enables businesses to obtain financial resources more easily and at a reduced cost (Xie et al. 2022). When economic stability increases energy demand and spurs economic expansion, this is known as the wealth effect (Shahbaz et al. 2022).

10.4 Methodology

10.4.1 Data

The data from 1999 to 2021 was used to examine the impact of natural resources, renewable energy, green technology, and green finance on financial growth in North African countries (Egypt, Algeria, Morocco, and Tunisia). The econometric model is as follows:

$$\mathrm{FD_{it}} = \psi_0 + \psi_1\mathrm{NR_{it}} + \psi_2\mathrm{RE_{it}} + \psi_3\mathrm{GT_{it}} + \psi_4\mathrm{GF_{it}} + \phi_{\mathrm{it}} \tag{10.1}$$

where financial development, natural resources, renewable energy, green technology, and green finance are represented by the variables $\mathrm{FD_{it}}$, $\mathrm{NR_{it}}$, $\mathrm{RE_{it}}$, $\mathrm{GT_{it}}$ and $\mathrm{GF_{it}}$ respectively. However, $\psi_0, \psi_1, \psi_2, \psi_3, \psi_4, i, t,$ and ϕ_{it} where it refers to the intercept. The coefficients for explanatory variables (NR, RE, GT and GF), sample of countries during (1999–2021), and white noise, respectively. In addition, Table 10.1 provides comprehensive details of the research variables.

10.4.2 Estimation Techniques

10.4.2.1 Cross-Sectional Dependence (CSD) Estimator

Our study employed the CSD test to identify the most effective statistical strategy. The proper unit root test technique (first or second generation) is determined in part by the test findings. Research findings may not be objective, consistent, or

Table 10.1 Variables measurements

Variables	Abb.	Measurement	Source
Financial development	FD	Financial development index	IMF (2025)
Natural resources	NR	Total natural resources rents (% of GDP)	WDI (2025)
Renewable energy	RE	(% of total final energy consumption)	WDI (2025)
Green technology	GT	The quantity of green technology innovation patents awarded for every 10,000 R&D employees	IRENA
Green finance	GF	International finance received for clean energy	OWD (2025)

dependable without CSD, leaving them open to bias, distortion, and limited use (Isiksal et al. 2024). To get around some of these issues, we employed CSD in our study. This is how the CSD equation is expressed:

$$\mathrm{CSD} = \left[\frac{\mathrm{TS}(N-1)_2^{1-\omega}}{2}\right]$$

where ω is equal to $\left(\frac{2}{N(N-1)}\right)\sum_i^N \sum_{j=i+1}^N \mu_{ij}$ where μ_{ij} is the correlation coefficients of the residuals of ADF estimation. Moreover the T represents time while the S represents cross-sectional area.

10.4.2.2 Slope Heterogeneity (SH) Estimator

Where γ and Δ are the "statistics utilised for estimating the slope heterogeneity". The coefficients of "pooled OLS and pooled weighted fixed effect" estimator are denoted by $\hat{\delta}$ and $\tilde{\delta}_{\mathrm{WFE}}$, respectively. Moreover, $Z_i W_T Z_i$ and K show the input matrix, identity matrix, an estimate of $\check{\theta}_i^2$, and predictor number, respectively. However, the adjusted slope heterogeneity is presented under:

$$\gamma = \sum_{i=1}^{N} (\hat{\delta} - \tilde{\delta}_{\mathrm{WFE}}) \frac{Z_i W_T Z_i}{\check{\theta}_i^2} (\hat{\delta} - \tilde{\delta}_{\mathrm{WFE}})$$

$$\check{\Delta} = \sqrt{M}\left(\frac{M^{-1}\check{X} - K}{\sqrt{2K}}\right)$$

where γ and Δ are the "statistics utilised for estimating the slope heterogeneity". The coefficients of "pooled OLS and pooled weighted fixed effect" estimator are denoted by $\hat{\delta}$ and $\tilde{\delta}_{\mathrm{WFE}}$, respectively. Moreover, $Z_i W_T Z_i$ and K show the input matrix, identity matrix, an estimate of $\check{\theta}_i^2$, and predictor number, respectively. However, the adjusted slope heterogeneity is presented under:

$$\check{\Delta}_{\mathrm{adj}} = \sqrt{M}\left(\frac{M^{-1}\check{X} - E(S_{\mathrm{iT}})}{\sqrt{\mathrm{VAR}(S_{\mathrm{iT}})}}\right)$$

10.4.2.3 Panel Unit Root Estimator

There may be some issues with the original stability strategy, according on evidence from the CD test findings. Therefore, we employed second-generation stability tests, namely the CADF and CIPS tests, to concentrate on the estimating inefficiencies. Equations below provide the mathematical expressions for these two tests:

$$\Delta\gamma_{\mathrm{it}} = \Delta\sigma_{\mathrm{it}} + \lambda_i W_{\mathrm{it}} + \varphi_{\mathrm{it}} + \sum_{j=1}^{n} \beta_{ij}\Delta W_{i,t-j} + \varepsilon_{\mathrm{it}}$$

$$\mathrm{CIPS} = M^{-1}\sum_{J=1}^{M} \mathrm{CADF}_J$$

where $\Delta\sigma_{\mathrm{it}}$, W_{it}, Δ, T and $\varepsilon_{\mathrm{it}}$ stand for "intercept, factor estimation, variance operator, interval, and white noise." The CIPS test is used to assess the permanent effects of factors on the fixed variables under study, with a particular focus on first differences.

10.4.2.4 Panel Cointegration Estimator

In panel cointegration testing, Westerlund (2007) suggests the null hypothesis, which is the lack of cointegration. Consequently, Westerlund's panel cointegration test does not impose any limitations on common factors and assumes error correction for both the panel as a whole members (S_t & S_a) and the panel as a whole (G_t & G_a).

10.4.2.5 Panel ARDL Approach

Pesaran et al. (1999) presented three models: the group mean (MG), the pooled group mean (PMG), and the dynamic fixed effect (DFE) model within maximum likelihood estimation (MLE). The following table shows the pooled panel regression in the error correction model within the PMG model:

$$\mathrm{FD}_{\mathrm{it}} = \mu_i + \sum_{s-1}^{n^1} \beta_{is}\mathrm{GF}_{\mathrm{it}-s} + \sum_{s-1}^{n^2} \vartheta_{ij}\mathrm{GT}_{\mathrm{it}-s} + \sum_{s-1}^{n^3} \varphi_{ij}\mathrm{RE}_{\mathrm{it}-s} + \sum_{S-0}^{p} \delta_{is}^{/}\mathrm{NR}_{\mathrm{it}-s} + \varepsilon_{\mathrm{it}}$$

where 1, 2, 3, …, N is the number of countries. t = 1, 2, 3, 4, …, t are the annual periods. (s) is the time lag. The lag of the independent variables is denoted by (p), while (n) is the lag of the independent variable. The main variables are financial development (FD), green finance (GF), green technology (GT), and renewable energy (RE), the vector of the control variable, i.e., natural resources, is NR, and finally ε_ is the error term with permanent effect. Pesaran et al. (1999) rearranged the formula as follows:

$$\Delta\mathrm{FD}_{\mathrm{it}} = \mu_i + \infty\mathrm{FD}_{2\mathrm{it}-1} + \dot{\lambda}_i X_{\mathrm{it}-1} + \sum_{s-1}^{n-1} \beta_{ij}\Delta\mathrm{FD}_{2\mathrm{it}-s} + \sum_{S=0}^{p-1} \dot{\delta}_{is}^{*}\Delta X_{\mathrm{it}-s} + \varepsilon_{\mathrm{it}}$$

where $\varphi = -1[1 - \sum_{s=1}^{p} \beta_{is}]$, $\lambda_i = \sum_{s=0}^{p} \delta_{is}$, $\beta_{is} = -\sum_{m=s+1}^{p}, \beta_{im}$, $j = 1, 2, \ldots, p-1$, and $\delta_{is} = -\sum_{m=s+1}^{p}, \beta_{im} j = 1, 2, \ldots, p-1$.

The error correction formula is employed to modify the following equation by parameters grouping the levels.

$$\Delta \mathrm{FD}_{2\mathrm{it}} = \mu_i + \infty_i\left(\Delta \mathrm{FD}_{2\mathrm{it}-1} - \dot{\theta}_i X_{it}\right) + \sum_{s=1}^{n-1} \beta_{is} \Delta \mathrm{FD}_{2\mathrm{it}-s} + \sum_{s=0}^{p-1} \delta_{is}^{*} \Delta X_{\mathrm{it}-s} + \varepsilon_{\mathrm{it}}$$

"Long-run equilibrium correlation among [$\Delta \mathrm{FD}_{2\mathrm{it}}$ and X_{it}] is [λ_i/υ_i]. The short-run coefficient is linked to past values and new growth dynamics, such as variations in [X_{it}] are indicated by [$\beta_{is}*$ and δ_{is}^{*}]. Finally, the error adjustment coefficient is [υ_i], which assesses the speed of adaptation of [$\mathrm{LIRE}_{\mathrm{it}}$] to the long run symmetry following a change in [X_{it}]". The factor or coefficient must be both significant and negative to build a long-run relationship ($\upsilon_i < 0$). Since the [υ_i] is significantly negative, it endorses integration. As a result, the following estimates are computed:

$$\hat{\gamma}_{\mathrm{PMG}} = \frac{\sum_{i=1}^{Q} \widetilde{\gamma_i}}{Q}, \quad \hat{\lambda}_{\mathrm{sPMG}} = \frac{\sum_{i=1}^{Q} \widetilde{\lambda_i}}{Q};$$

$$\hat{\varphi}_{\mathrm{sPMG}} = \frac{\sum_{i=1}^{Q} \widetilde{\varphi_i}}{Q} \quad \text{and} \quad \hat{\varpi}_{\mathrm{sPMG}} = \frac{\sum_{i=1}^{Q} \widetilde{\varpi_i}}{Q}$$

where $s = 0, \ldots, p-1$. $\hat{\gamma}_{\mathrm{PMG}} = \tilde{\gamma}$.

The model below is derived based on the approach of Eq. 5 (equation above the previous one):

$$\Delta \mathrm{FD}_{i-t} = \mu_i + \infty_i(\Delta \mathrm{FD}_{2i,t-1} - \beta_1 \mathrm{GF}_{i,t-1} \beta_2 \mathrm{GT}_{i,t-1} \beta_3 \mathrm{RE}_{i,t-1} \beta_4 \mathrm{NR}_{i,t-1}) + \sum_{s-1}^{n-1} \varpi_s^i \Delta(\Delta \mathrm{FD}_{2\mathrm{it}})_{t-s} + \sum_{s=0}^{n-1} \vartheta_{1s}^i \Delta(\Delta \mathrm{GF}_{\mathrm{it}})_{t-s} + \sum_{s=0}^{n-1} \vartheta_{2s}^i \Delta(\Delta \mathrm{GT}_{\mathrm{it}})_{t-s}$$

The PMG estimator takes short-term variation into account while assuming long-term homogeneity. The next estimator after the panel ARDL approach is the MG estimator, which is employed in country-level research. In contrast, the PMG strategy necessitates both short-term and long-term variance. The estimator's consistency and dependability depend on the size of the series. The DFE estimator is the last strategy for the ARDL technique. Both short-term and long-term homogeneity restrictions are used in the DFE method.

10.5 Outcomes and Discussions

10.5.1 Descriptive Test Results

The results of descriptive statistics are shown in Table 10.2. The results of the average values show the central tendency of the studied variables, where the values of NR, RE, GT, FD, and GF were 9.734, 8.463, 2.263, 0.241, and 1.56, respectively. The values of variance or dispersion for the variables (NR, RE, GT, FD, and GF) were 9.263, 5.926, 2.931, 0.086, and 3.33, respectively.

Furthermore, the preliminary test results are shown in Table 10.3. The t-statistics value results of the CSD test for FD (2.743), GF (0.425), GT (3.943), RE (7.876), and NR (7.132) are significant at the 1% significance level, indicating the presence of CSD.

The stationarity properties of the series were verified using the CIPS and CADF tests by Pesaran (2007) as shown in Table 10.4. The results showed that none of the pivotal variables accepted the non-stationarity hypothesis at this level. The results confirmed the mixed order of stationarity $I(0)$ and $I(2)$.

Table 10.5 presents the results of the Westerlund cointegration test. The null hypothesis of no cointegration among the FD, GF, GT, RE, and NR statistics was rejected (with the exception of Ga and Pa, which may be due to the small sample size). These reports can be used as evidence of cointegration for the entire panel or for a single country within its panels.

Table 10.2 Descriptive outcome

Variable	Obs.	Mean	Std. Dev.	Min	Max
NR	92	9.734388	9.262695	0.194633	34.19132
RE	92	8.463043	5.9258	0.1	22.4
GT	92	2.26268	2.931058	0	13
FD	92	0.240789	0.086907	0.113803	0.434173
GF	92	1.56E+08	3.33E+08	0	2.06E+09

Table 10.3 CSD outcome

Variable	CD-test	*p*-value
FD	2.743	0.006*
GF	0.425	0.036*
GT	3.943	0.00*
RE	7.876	0.00*
NR	7.132	0.00*

* Significant statistic

Table 10.4 Stationary outcome

Variables	Level		First difference	
	CIPS	CADF	CIPS	CADF
FD	– 2.17	– 0.610 [0.27]	– 4.053*	– 3.617 [0.00]
GF	– 3.936*	– 2.046 [0.02]*	– 5.334*	– 3.366 [0.00]
GT	– 2.839*	– 2.186 [0.01]*	– 4.398*	– 4.458 [0.00]
RE	– 2.896*	– 0.652 [0.25]	– 5.093*	– 3.130 [0.00]
NR	– 1.576	– 1.094 [0.13]	– 3.318*	– 1.868 [0.03]

* Significant statistic

Table 10.5 Cointegration outcome

Statistical	Value	*Z*-value	*p*-value
Gt	3.061	9.801	0.00*
Ga	0.06	2.795	0.997
Pt	1.422	3.674	0.00*
Pa	0.047	1.685	0.954

* Significant statistic

10.5.2 Panel ARDL Results

According to Table 10.6's results from the MG, PMG, and DFE estimations, all of the independent variables that were chosen—green finance (GF), green technology (GT), renewable energy (RE), and natural resources (NR)—have statistically significant long- and short-term effects on financial development (FD), though the extent and direction of these effects differ depending on the estimator. These findings show consistency and support the idea that contributions to sustainable development are essential in determining the financial environment of North African nations.

The information shows that the rates at which the financial systems in various nations transition from short-term imbalance to long-term equilibrium differ. Notably, the PMG model yielded an error correction term (ECT) of – 0.3738, which is statistically significant at the 1% level. This means that throughout each period, around 37.38% of the deviations from the long-term path are corrected.

This demonstrates that the variables have a steady, long-term connection. All of the MG, PMG, and DFE estimators' long-run coefficients for green finance (GF) are positive and statistically significant at the 5% level. In particular, the PMG estimator shows that a little but noteworthy improvement in financial development is linked to a 1% rise in GF (coefficient = 4.8E−11). Similarly, in the short run, a 1% increase in ΔGF leads to an increase in financial development. Likewise, a 1% rise in ΔGF boosts financial development (DFE) by 1.51E−11 in the near term, indicating that greater access to and investment in green finance has a favorable, both immediate and long-term, influence on the financial system of North

Table 10.6 MG, PMG and DFE outcome

Verb	MG		PMG		DFE	
	Coe	*p*-value	Coe	*p*-value	Coe	*p*-value
Long term						
GF	1.17E−09	0.012	4.8E−11	0.044	6.90e−11	0.037
GT	0.00284	0.027	0.00449	0.01	0.0001	0.019
RE	− 0.0204	0.01	0.00143	0	0.0138	0.003
NR	0.00318	0.001	0.00034	0.573	0.0011	0.048
Short term						
ΔGF	5.88e−10	0.305	1.6E−10	0.253	1.51e−11	0.034
ΔGT	0.00062	0.046	0.00114	0	0.0006	0.048
ΔRE	0.00092	0.006	0.00342	0.476	0.0044	0.046
ΔNR	0.00246	0.063	0.00183	0.015	0.0006	0.468
ECT(−1)	− 0.2995	0.005	− 0.37383	0	− 0.203	0.033

* Significant statistic

African nations. These findings align with those proposed by Ali et al. (2024). Additionally, throughout both time periods, there is a substantial and statistically significant positive correlation between financial development and green technology (GT). According to the long-run PMG estimate (0.0045, $p = 0.01$), a 1.0% rise in GT eventually results in a 0.0045% increase in FD. A 1.0% rise in ΔGT in the short term raises FD by 0.0006 (DFE), supporting the idea that advancements in green technology help the growth of the financial system. The findings align with research conducted by Abidin et al. (2021) and Hwang et al. (2022). Despite variations between models, the impact of renewable energy (RE) is finally determined to be favorable and considerable in both the PMG and DFE estimations. FD rises by 0.0014% over time with every 1.0% increase in the utilization of renewable energy, according to the PMG calculation (0.0014, $p = 0.00$). Likewise, the short-term projection (0.00092 in MG, 0.0044 in DFE) highlights the switch to renewable energy's immediate financial benefits. These findings bolster the claim that funding clean energy improves financial markets while simultaneously helping the environment. The study's findings corroborate those of Chireshe (2021) and Prempeh (2023).

There is also a beneficial correlation between financial development and natural resources (NR). According to the statistically significant long-term MG and DFE estimations, when NR increases by 1.0%, FD increases by 0.003% and 0.0011%, respectively. But according to the PMG model, there is some variation in the long-term effects of resource wealth, since the long-term association is statistically insignificant ($p = 0.573$). The MG and PMG models show a stronger effect in the short term (0.00246 and 0.00183, respectively), while the DFE model does not show a statistically significant effect. This suggests that natural resource rents

Table 10.7 Causality outcomes

Null hypothesis	*F*-statistic	Prob.	Causality direction
NR ≠ FD	2.0169	0.0437	→
FD ≠ NR	1.3118	0.1896	
RE ≠ FD	5.5346	0.0000	↔
FD ≠ RE	3.9775	0.0001	
GF ≠ FD	– 0.4400	0.4500	←
FD ≠ GF	– 0.5500	0.0060	
GT ≠ FD	– 0.7536	0.0010	→
FD ≠ GT	– 0.7514	0.6460	

could give the financial sector immediate liquidity, though this effect might eventually fade. Asif et al. (2020), Oben (2022), and Sun et al. (2020) are supported by the study's findings.

To improve the quality of the empirical data, this study additionally employed paired Granger causality analysis to ascertain the direction of causal linkages between variables, as indicated in Table 10.7. In summary, the study discovered one bidirectional causal link and three unidirectional ones. Particularly, it was shown that there was unidirectional causation from GF to FD, from GT to FD, and from NR and FD. Concurrently, it was shown that RE and FD are causally related in both directions.

10.6 Discussion

This study examined the influence of green finance, green technologies, renewable energy, and natural resources on the financial development of North African nations. The findings from the MG, PMG, and DFE estimators unequivocally support the hypothesis that sustainability-oriented inputs influence financial growth in the region, both in the short and long term, albeit with differing intensity and direction contingent upon the model specification employed. The residual value of the adjustment was confirmed by the existence of a statistically significant negative error correction term (–0.3738) derived from the PMG estimator. This signifies the presence of a long-term equilibrium among the examined variables and implies that around 37.38% of the short-term disequilibrium is rectified each period. Consequently, it may be contended that North African nations possess diverse financial systems; yet, they generally demonstrate a moderate rate of adjustment towards long-term equilibrium when influenced by sustainability policies and practices.

Green Finance (GF) exhibits the most significant restricting impact on financial growth in NDGF, indicating that increased investment in green-directed funding correlates with enhanced financial development in North Africa. These results are particularly significant for Morocco and Egypt, as both nations have just formulated their green finance strategy and commenced the issuance of green bonds. The immediate and enduring advantages indicate that these North African nations aim to enhance their financial markets and access to sustainable financing, while also striving to improve investor perception, diversify capital inflows, and facilitate lasting transformations in the economic structure concerning climate objectives.

The influence of Green Technology (GT) activities on financial development was significant and beneficial, reinforcing the assertion that progress in sustainable sectors contributes to financial advancement. The significance of this pertains to Algeria and Tunisia, where the cleantech and green innovation sectors have advanced through government subsidies and private sector support. Further advancements can mitigate long-term risk while enhancing and attracting investment through greater productivity and energy efficiency, thereby strengthening the country's financial framework.

The effects of Renewable Energy (RE) were quite intricate, demonstrating positive and significant impacts in both PMG and DFE calculations. The rise of renewable energy, especially solar and wind resources prevalent in North Africa, both internationally and regionally, contributes to the expansion of the financial sector by attracting investment, reducing reliance on energy imports, and fostering innovative financing mechanisms associated with clean energy infrastructure. Egypt and Morocco have notably excelled in the advancement of renewable energy, and the financial advantages shown in this study underscore the significance of investment developments in these nations. A reinforcing loop exists between renewable energy (RE) and financial development, whereby each sustains the other; increased investment in renewable energy fosters financial development, which subsequently promotes additional investment in energy.

Natural Resources (NR), while traditionally a strength for North African economies like Algeria and Libya, fund development in these sectors; nonetheless, the long-term benefit was negligible under the PMG model. This indicates that although natural resource rents temporarily infuse capital into the financial system, particularly in extractive industry nations, sustainable long-term growth is contingent upon effective management and governance of these resources, along with diversification and investment in advanced industries.

The pairwise Granger causality findings enhance comprehension of these interactions. The evident unidirectional causality from GF, GT, and NR to financial development indicates that these variables serve as precursors or influencers rather than results. This indicates the necessity for additional proactive investment and policy reformation in these domains. The bidirectional causation between renewable energy (RE) and financial development (FD) illustrates a mutually reinforcing link, wherein clean energy investment and the expansion of the financial system enhance and fortify one another in a self-sustaining manner.

10.7 Conclusion and Policy Recommendation

10.7.1 Conclusion

The study examined the impacts of green finance, green technology, renewable energy, and natural resources in North African nations, utilising the MG, PMG, and DFE estimators. The results from all estimating methods employed in this research indicate that all four variables exert a beneficial influence on the level of financial development, both in the short and long term, but to varying degrees that are statistically significant. A steady long-run equilibrium relationship, supported by a substantial error correction term, demonstrates the sustainable development factors in the region and the adaptive evolution of the financial system.

The evolution of green finance reflects the significance of green investment instruments in fostering the growth of financial markets, as it has become a fundamental catalyst for financial advancement. Similarly, the impacts of green technology and renewable energy were determined to be markedly beneficial, highlighting the strategic importance of such innovations in extensive economic diversification and sustainable capital flow contributions. Despite the uneven impact of natural resources, the short-term results confirm that resource richness is crucial for liquidity provision and the robustness of the financial sector, especially in resource-dependent nations such as Algeria and Libya.

The causality analysis validated the robustness of these connections, indicating that green finance, green technology, and natural resources exerted a unidirectional causal influence on financial development, whereas the association with renewable energy was bidirectional. These findings underscore an alternative perspective in North Africa, wherein sustainability is treated as an independent variable rather than a dependent outcome of financial development.

10.7.2 Policy Recommendations

Based on the outcomes of the investigation, the following policy recommendations are proposed to enhance financial development through sustainable pathways in North African nations:

- Enhance Green Finance Frameworks: Governments ought to formulate and implement policies for green finance, including directions for tax incentives, regulations, disclosures for climate bonds, green investment accounts, ESG-related investments, and other financial instruments. The Capital Market Authority of Morocco and other institutions could serve as a paradigm for the developing world in the advancement of green financial markets.
- Prioritise investment in emerging green technologies: Allocating resources to the research and development of green technology must be paramount. Scaling Environmentally Sustainable Technologies (SEST), financed through public–private partnerships and various clean technology innovation centres,

contributes to environmental protection and the enhancement of employment opportunities.

- Advocate for the integration of renewable energy: The expansion of renewable energy, particularly solar and wind technologies, must remain a primary focus in the nation's development strategies. All accessible renewable energy resources should be employed to stimulate activity in the financial industry and diminish dependence on fossil fuels. The financing of renewable energy infrastructure should be sourced from green international funds and public–private partnerships (PPP).
- Supervise natural resources: The administration of revenues from natural resources must be transparent and efficient in resource-rich nations. Establishing sovereign wealth funds, improving fiscal transparency, and channeling resource rents into productive and diversified sectors will enhance long-term financial stability.
- Foster Regional Collaboration: North African nations encounter a collective challenge of disjointed shared learning and policy formulation in domains such as sustainable finance, green innovation, and energy transformation. Regional platforms would enable the transmission of investment knowledge, foster collaboration on investment initiatives, and consolidate standards.
- Promote Financial Inclusion and Literacy: Comprehensive development policies must prioritise enhanced access to finance for marginalised communities and small to medium firms, in conjunction with sustainable public investment that improves ecosystem health. Awareness campaigns for green finance and sustainable investment can strengthen the structure of the developing green financial ecosystem.

Incorporating North African concerns into environmental financial development promotes inclusive growth, while climate mitigation initiatives enhance intergenerational fairness and social results. The recommended approaches will facilitate global efforts in climate change mitigation while improving financial performance.

References

Abidin Z, Prabantarikso RM, Wardhani RA, Endri E (2021) Analysis of bank efficiency between conventional banks and regional development banks in Indonesia. J Asian Finance Econ Bus 8(1):741–750

Ali M, Igunnu PO, Tursoy T (2024) Do green finance and energy prices unlock environmental sustainability in Pakistan? Fresh evidence from RALS-EG cointegration. OPEC Energy Rev

Asif M, Khan KB, Anser MK, Nassani AA, Abro MMQ, Zaman K (2020) Dynamic interaction between financial development and natural resources: evaluating the 'Resource curse' hypothesis. Resour Policy 65:101566

Awwad BS (2022) The role of environmental disclosure in enhancing the market value of industrial and service companies listed on the Palestinian stock exchange. Int J Entrepr 26(2):1–12. ISSN: 1939-4675

Awwad B (2025a) Climate finance and sustainability: a story told through the Journal of Cleaner Production. EuroMid J Bus Tech-Innov 31–47

Awwad BS (ed) (2025b) Sustainability in light of governance and artificial intelligence applications. Emerald Group Publishing
Awwad BSAL (2024) Governance with relationship between entrepreneurship and economic growth in Palestine. Int J Law Manag 66(2):259–287
Bhammar H, Li W, Molina CMM, Hickey V, Pendry J, Narain U (2021) Framework for sustainable recovery of tourism in protected areas. Sustainability 13(5):2798
Chireshe J (2021) Finance and renewable energy development nexus: evidence from Sub-Saharan Africa. Int J Energy Econ Policy 11(1):318–325
Dimnwobi SK, Madichie CV, Ekesiobi C, Asongu SA (2022) Financial development and renewable energy consumption in Nigeria. Renew Energy 192:668–677
Hwang J, Abbas J, Joo K, Choo S-W, Hyun SS (2022) The effects of types of service providers on experience economy, brand attitude, and brand loyalty in the restaurant industry. Int J Environ Res Public Health 19(6):3430
IMF (2025) International Monetary Fund Data. https://data.imf.org/?sk=f8032e80-b36c-43b1-ac26-493c5b1cd33b
Isiksal AZ, Assi AF, Rakhmetullina SZ, Zhakanov A (2024) Sustainable environment and military expenses: the role of energy innovations. J Environ Account Manage 12(03):299–310
Khan MA, Gu L, Khan MA, Oláh J (2020a) Natural resources and financial development: the role of institutional quality. J Multinatl Financ Manag 56:100641
Khan MA, Popp J, Talib MNA, Lakner Z, Khan MA, Oláh J (2020b) Asymmetric impact of institutional quality on tourism inflows among selected Asian Pacific countries. Sustainability 12(3):1223
Khan MK, Babar SF, Oryani B, Dagar V, Rehman A, Zakari A, Khan MO (2022) Role of financial development, environmental-related technologies, research and development, energy intensity, natural resource depletion, and temperature in sustainable environment in Canada. Environ Sci Pollut Res 29:622–638
Mlachila M, Ouedraogo R (2020) Financial development curse in resource-rich countries: the role of commodity price shocks. Q Rev Econ Finance 76:84–96
Nawaz MA, Seshadri U, Kumar P, Aqdas R, Patwary AK, Riaz M (2021) Nexus between green finance and climate change mitigation in N-11 and BRICS countries: empirical estimation through difference in differences (DID) approach. Environ Sci Pollut Res 28:6504–6519
Oben RJ (2022) The impact of natural resources on financial development: the global perspective Eastern Mediterranean University (EMU)-Doğu Akdeniz Üniversitesi (DAÜ)
OWD (2025) International finance received for clean energy. https://ourworldindata.org/grapher/international-finance-clean-energy
Pesaran MH (2007) A simple panel unit root test in the presence of cross-section dependence. J Appl Economet 22(2):265–312
Pesaran MH, Shin Y, Smith RP (1999) Pooled mean group estimation of dynamic heterogeneous panels. J Am Stat Assoc 94(446):621–634
Prempeh KB (2023) The impact of financial development on renewable energy consumption: new insights from Ghana. Future Bus J 9(1):6
Rezagholizadeh M, Abdi Y (2022) Financial development and development of renewable energy technologies: a comparison of developing and developed countries. Economics and Policy of Energy and the Environment (2022/1)
Saygin O, Iskenderoglu O (2022) The nexus between financial development and renewable energy consumption: a review for emerging countries. Environ Sci Pollut Res 29(10):14522–14533
Shahbaz M, Aboutorabi MA, Yazdi FA (2024) Foreign capital, natural resource rents and financial development: a new approach. Glob Bus Rev 25(2):401–420
Shahbaz M, Nasir MA, Lahiani A (2022) Role of financial development in economic growth in the light of asymmetric effects and financial efficiency. Int J Financ Econ 27(1):361–383
Skare M, Gavurova B, Sinkovic D (2023) Regional aspects of financial development and renewable energy: a cross-sectional study in 214 countries. Econ Anal Policy 78:1142–1157
Sun Y, Ak A, Serener B, Xiong D (2020) Natural resource abundance and financial development: a case study of emerging seven (E-7) economies. Resour Policy 67:101660

WDI (2025) World development indicators. https://databank.worldbank.org/source/world-development-indicators
Westerlund J (2007) Testing for error correction in panel data. Oxford Bull Econ Stat 69(6):709–748
Xie Z, Liu X, Najam H, Fu Q, Abbas J, Comite U, Cismas LM, Miculescu A (2022) Achieving financial sustainability through revenue diversification: a green pathway for financial institutions in Asia. Sustainability 14(6):3512
Yıldırım S, Gedikli A, Erdoğan S, Yıldırım DÇ (2020) Natural resources rents-financial development nexus: evidence from sixteen developing countries. Resour Policy 68:101705
Zhang C, Khan I, Dagar V, Saeed A, Zafar MW (2022) Environmental impact of information and communication technology: unveiling the role of education in developing countries. Technol Forecast Soc Chang 178:121570

Greenwashing and the Need for International ESG Standards: A Comparative Analysis of EU and US Regulation

11

Mira Moubarak, Razan Awadallah Awwad, and Tala Awwad

Abstract

The trend of sustainable investing and corporate social responsibility has increased demand for Environmental, Social, and Governance (ESG) disclosures. While this has resulted in an increase in disclosures, Greenwashing has also become a problem. When an issuer's publicly stated sustainability profile and the features and/or goals of a financial instrument or financial product, either by action or omission, do not accurately reflect the underlying sustainability risks and impacts associated with that issuer, financial instrument, or financial product, this is known as "greenwashing". Despite efforts to prevent greenwashing, such as regulations in the European Union, the problem persists worldwide. This paper will thoroughly study this topic through the lens of ESG disclosure regulations. It compares the situation of ESG disclosures with greenwashing in the Europe Union with the United States of America and highlights the challenges and needs for more effective measures to ensure accurate and transparent ESG disclosures and aims to prevent greenwashing. It proposes that by comparing different approaches, policymakers can identify the strengths and weaknesses of each while using this information to develop an internationally standardized, comprehensive, and effective framework for ESG disclosure regulations.

M. Moubarak
International and Comparative Law, University of Zurich, Zurich, Switzerland

R. A. Awwad (✉)
Business and Economics, Al-Quds University, Jerusalem, Palestine
e-mail: Razan.awadallah@gmail.com

T. Awwad
Department of Law, Birzeit University, Birzeit, Palestine

B. Awwad and R. A. Awwad (eds.), *Technology and Innovation Towards a Future of Business Sustainability*, Contributions to Environmental Sciences & Innovative Business Technology, https://doi.org/10.1007/978-981-95-6469-9_11

Keywords

ESG • Regulation • Europe • United States of America • Greenwashing

11.1 Introduction

In the last decade, a growing global awareness regarding sustainability was apparent. The worldwide Sustainable Investment Alliance (GSIA 2020) reports that at the beginning of 2020, worldwide sustainable investment assets were $35.3 trillion, a 15% rise from 2018. This amount represents more than a third of all professionally managed assets globally, and Europe accounts for the largest share of sustainable investing assets, followed by the USA. Global.

Subsequently, incorporating ESG factors into business operations and reporting procedures has become crucial for the capital market, investors, and regulators. The integration has caused corporate social responsibility and ESG disclosure regulations to gain massive legislative push. This is consistent with findings from emerging markets such as Palestine, where environmental disclosure has been shown to positively impact firm market value (Awwad 2022). The EU is the leading region with the most advanced regulations, followed by the USA regarding legislation and momentum with growth initiatives (Eccles and Klimenko 2019; La Torre et al. 2020). However, this momentum requires a pause on what governments and the private sector can accomplish through the disclosures.

Despite some benefits derived from specific policies, the desired level of success for regulations has yet to be attained. While it would be unfair to solely blame governments and corporations for this predicament, it is evident that there is an incomplete implementation of ESG policies and a lack of successful integration into corporate policies. Consequently, ESG disclosures face numerous challenges and risks, with greenwashing being one of the most concerning threats (Delmas and Burbano 2011; European Securities & Markets Authority [ESMA] 2021). The notion of greenwashing is broad and various definitions can be found in other literature. Still, this study will adopt the most used and advanced definition put forth by the European Securities and Markets Authority. This definition avoids limiting ESG factors to only environmental issues and does not restrict greenwashing to a specific type (ESMA 2021).

Notwithstanding several national ESG regulations and voluntary international recommendations, the persisting challenges in the market highlight the significance of moving away from the current confusing legislative landscape and towards establishing internationally standardized ESG disclosure regulations (Sullivan and Mackenzie 2020; Awwad 2025a, b).

The absence of a consensus regarding ESG disclosure regulations and greenwashing in the broader context, coupled with a lack of comparative research between the EU and US markets and a holistic examination of the regulations, has necessitated the initiation of this study. The paper primarily aims to scrutinize the inadequacies of ESG regulations in the USA and EU and assess their impact on greenwashing to create an opportunity for the future of international standardized ESG disclosure regulations.

The study aims to address the gap in greenwashing literature in three ways. Firstly, it adopts a comprehensive approach to examining ESG regulations, in contrast to other literature that focuses separately on E, S, or G issues. Secondly, the study examines the complex relationship between ESG disclosures and greenwashing, as well as the impact of the former on the latter. Thirdly, by conducting a comparative analysis of the approaches taken by the EU and the USA, the study aims to promote the standardization of regulations.

11.2 Literature Review

The primary emphasis of the study is on two crucial aspects: the comparison of ESG disclosure regulations between the European Union and the US and their correlation with the occurrence of greenwashing.

11.2.1 Comparing the EU to the USA

There have been several pieces of literature on the importance of performing a comparative study to examine sustainability and ESG. Some studies focused on examining global samples of companies from diverse countries, aiming to address multiple research gaps and respond to calls for further investigation into ESG disclosures, as identified in prior studies (Doan and Sassen 2020; Mateo-Márquez et al. 2022; Yu et al. 2020). Other literature tackled the topic through a comparative study building on previous research that conducts a regional comparison to assess the influence of sustainability and climate change-related pressures on corporate greenwashing behavior.

The study supports the idea of international standardized regulation; therefore, it supports the opinion that governmental legislation is the primary source of pressure institutional regulatory frame, which compels organizations to comply with certain behaviors through formal sanctions (Bruton et al. 2010). Furthermore, regulations can affect the conduct of companies outside their jurisdiction, as the laws of a country establish social norms for organizations operating within its borders (Mateo-Márquez et al. 2022). Also, national legislation is critical in reducing environmental harm, as it raises the abilities of international agreements and makes their implementation smoother, according to Townshend (2014).

ESG originated in developed countries such as Europe and the USA and has reached a relatively mature stage of practice and development within companies. Several pieces of literature have explored the issue of greenwashing, offering unique insights into the phenomenon and highlighting the importance of global cooperation in addressing environmental challenges.

A business is connected to an active environment with several opportunities and risks. A country's economic, social, and environmental model has a notable influence on business practices (Ortas et al. 2015), but it indirectly alters the approach of investors, who then accelerate the means of ESG implementation by businesses (Clementino and Perkins 2020). Past, present, and future events, like economic depression, war, and epidemics, resulted and will result in situations of severe emergency in the economic system of countries (Crilly et al. 2016). This sheds light on a country's unregulated markets, emphasizing the importance of standardization and social responsibility. The COVID-19 pandemic has amplified enterprises' environmental risk exposure in developing economies, a concern previously primarily pertinent to developed economies.

These risks have had the effect of impeding the environmental performance of businesses or accelerating the process of sustainable development in developing nations.

A number of works have been published regarding the influence of legislators on greenwashing and express varying opinions on the impact of a country's regulations on businesses greenwashing (Cho et al. 2015; Roulet and Touboul 2015; Haque and Ntim 2018). Regulators and policymakers must conduct extensive research on greenwashing and corporate social responsibility disclosures to generate regulations that prevent organizations from engaging in greenwashing practices (Delmas and Burbano 2011).

These divergent viewpoints suggest that further research is needed, particularly studies evaluating extensive data by country to facilitate cross-country comparisons (Lyon and Montgomery 2015). This interplay between governance and sustainable development is also evident in emerging economies, where entrepreneurial frameworks are closely tied to governance structures (Awwad 2024).

11.2.2 ESG Disclosure Regulations

The debate around the relationship between corporate social responsibility, firm value, and performance has been ongoing since Friedman's argument favoring unrestrained capitalism and against any activity that may distort economic freedom, including socially responsible actions. Despite over 50 years of research, this topic continues to generate a wide range of discussions and diverse viewpoints, but no consensus has been reached. Studies have taken different approaches to ESG, reflecting various perspectives and definitions. Contrary to Friedman, CSR practices enhance the firm's profit and reputation (Boubaker et al. 2022). There has been mounting evidence over the years that adopting an ESG approach to operations can demonstrate profit for companies and investors (Miralles-Quirós et al. 2019), and embracing corporate social responsibility has improved financial performance according to Flammer (2014). Investors are attracted to companies with an ESG approach incorporated into their operations (Hartzmark and Sussman 2019). Recent studies in the MENA region support these trends, demonstrating that board diversity and IT governance also significantly influence both financial performance and ESG-related outcomes (Awwad et al. 2023a, b; Awwad and El Khoury 2024). Investors commonly use ESG as a standard and approach to assessing corporate behavior and future financial performance. The assessment of the sustainable development of enterprises, consider the three fundamental ESG factors critical for investment decision-making and analysis. Responsible investment is defined by the (PRI) as "a strategy that incorporates ESG factors into investment decisions and active ownership."

11.2.3 Greenwashing

Although some individuals perceive greenwashing as a novel concept that only emerged in recent years, it has been a topic of discussion for some time. However, it has gained more legitimacy over time, and its effects are now being more widely recognized; "sharp increase in articles since 2011" on greenwashing literature (Lyon and Montgomery 2015). Additionally, "references to greenwashing in the literature have rapidly increased, and the types and consequences of greenwashing have become a research hotspot" (Zhang et al. 2023). The evolution in academic discourse and literature linked to greenwashing has been substantial (Pope and Wæraas 2015).

11.2.4 ESG Disclosure in Relation to Greenwashing

In the context of ESG disclosures, greenwashing threatens the reliability and accuracy of reporting. The characterization of greenwashing is mainly the difference between what a company prefers to disclose regarding its ESG performance and its actual conduct relating to such factors. This idea is conveyed as the "incongruence between the reputational intention and the actual, real sustainability performance of the company." Hence, greenwashing damages the interest of all market actors (Steiner et al. 2018). Greenwashing is a form of communication that leads people, including consumers and stakeholders, to develop a distorted view of an organization's environmental performance and benefits (Tateishi 2017). In contrast, greenwashing can be more expansive, incorporating social aspects by referring to "the selective disclosure of positive information about a company's environmental or social performance without disclosing negative information, which creates an excessively positive corporate image (Lyon and Maxwell 2011)."

Greenwashing was also defined as "the collision of two different firm behaviors, which occur when firms have strong positive communication about their environmental contribution while still having poor environmental performance." This allows unethical behaviors to gain significant financial profit, influence outcomes, and solicit trust (Delmas and Burbano 2011). It can be "an umbrella term for various misleading communications and practices that intentionally or not induce false positive perceptions of an organization's environmental performance."

Earlier research has concentrated on defining greenwashing, with some studies defining it as positive communication about environmental performance despite having poor actual performance. Lyon and Maxwell (2011) expanded the definition by adding social aspects, where companies disclose positive information while ignoring negative information to create a positive image. Other studies have classified greenwashing into several types, including selective disclosure (De Freitas Netto et al. 2020; Uyar et al. 2020). These classifications also encompass decoupling, greenwashing (Berrone et al. 2015), and attention deflection, which refers to practices that divert stakeholders' focus away from negative environmental impacts (Yu et al. 2020). The relationship between ESG and greenwashing was noted that greenwashing harms market actors (Steiner et al. 2018). While it "is an umbrella term for various misleading communications and practices that intentionally or unintentionally create false positive perceptions of an organization's environmental performance" (Lyon and Maxwell 2011).

Regarding the impact of national regulation and comparing the different regions and its influence on ESG and greenwashing, prior literature has that a country's legal landscape can affect business practices. Another argued that it indirectly transforms the mindset and can affect ESG adoption. Conducting comparative research by examining a global sample and studying different regions is supported by Yu et al. (2020). As mentioned before, these divergent viewpoints suggest that the countries need further research to facilitate cross-country comparisons (Lyon and Montgomery 2015).

This paper presents a comparative study of ESG regulations in the USA and EU, focusing on their relation to greenwashing. It will disclose the inadequacies of current ESG regulations and highlight the need for international standardized ESG disclosure regulations in the future. The study aims to prove that the current confusion of legislation regarding ESG disclosures is hindering progress toward sustainability, and international standardized regulations will help companies to implement ESG policies more effectively while reducing the risk of greenwashing.

11.3 Research Methodology

This paper employs a comparative research method to analyze and compare the ESG regulations and greenwashing phenomenon between two regions, namely the United States and the European Union. The research adopts a qualitative approach and relies on collecting qualitative data. Specifically, the study examines disclosure regulations related to ESG currently in place in both regions from 2010 to 2023, as well as international and voluntary initiatives. The rationale behind this is to gain insight into the regulations in the top financial markets and explore their similarities and differences regarding greenwashing.

To ensure comprehensive research, the study focuses on academic research and screens articles published in academic journals between 2010 and 2023 that contain the search term greenwash or greenwashing, ESG or disclosure regulations, and common alternatives. The study is not limited by subject area due to the cross-region nature of the topic. However, it excludes irrelevant articles such as those appearing due to author names. The study mainly focuses on studies that address greenwashing concerning ESG scores and performance while excluding studies that deal with greenwashing as misleading advertising for consumer products. The collected data includes ESG disclosure regulations, enforcement mechanisms, greenwashing practices, challenges, and improvement strategies. The study aims to conduct a comparative analysis of ESG disclosure regulations to assess their effectiveness and examine factors contributing to the prevalence or deterrence of greenwashing. The study's findings will contribute to developing more effective policies and practices at the national level, with a primary focus on the importance of international development.

11.4 Findings and Discussion

11.4.1 Europe ESG Toolkit

ESG importance has grown significantly over time with the increase of environmental awareness, leading to social and governance awareness later, and Europe is leading and the most advanced regarding the topic.

Furthermore, there has been a substantial increase in corporate adoption of enhanced governance practices and sustainability trends, leading to a surge in the availability of ESG information for investors. Notably, the EU has been at the forefront of adapting to the evolving market dynamics and demands. Accordingly, it introduced a comprehensive framework of standards to govern ESG considerations, encompassing regulations related to disclosure and taxonomy. This regulatory activity has intended to adopt ESG investment to encourage investors to integrate ESG facts in their investment decision process, to control the needed accordance of corporations with the ESG, and mainly to meet the Paris Accord milestone.

This reflects the findings in other jurisdictions where robust governance frameworks significantly contribute to ESG transparency and digital transformation (Awwad et al. 2023a, b; Awwad et al. 2024).

Greenwashing in Europe. This section investigates the phenomenon of greenwashing in the European Union. The discussion will launch with an overview of greenwashing, followed by examining the triggering factors contributing to its prevalence. A review of past and ongoing greenwashing cases is conducted to provide a comprehensive perspective. Finally, this section will consider the preventive measures that were and will be adopted to counteract greenwashing practices.

The difficulty of the greenwashing phenomenon in Europe can be illustrated by the European Supervisory Authorities (ESA) "Call for Evidence." The EC asked the ESAs to provide their views on "the risks and occurrences of greenwashing in the EU financial sector, the supervisory actions taken, and the challenges in addressing these risks." The goal is to define and understand the scale and potential risks of greenwashing and issue recommendations based on the findings. ESA is "seeking input on potential greenwashing practices across the entire financial sector of the EU, encompassing banking, insurance, and financial markets, which may be relevant to various segments of the sustainable investment value chain and the financial product lifecycle."

The call is comprehensive, covering all aspects of the ESG fields, with the EC using the term broadly to refer to sustainability-related claims. The ESAs identify potential greenwashing features, and the call seeks evidence at both entity and product levels, within and outside the scope of EU sustainable finance legislation. The call acknowledges increased misleading environmental claims but not necessarily in governance and social aspects. It aims to collect information in the sectors under the three ESAs. The "Call" indicates a significant prevalence and severity of greenwashing cases in Europe, with an increase in their numbers. The call underscores the necessity for new regulations and rules to tackle this challenging issue in the market.

USA ESG Toolkit. Under this section, the paper will give an overview of present ESG disclosure regulations in the USA and greenwashing, including cases and preventive measures.

The increasing focus on ESG in the USA reflects a growing recognition that companies must take a complete attitude to manage their performance and that sustainability, social responsibility, and good governance are essential to long-term success. Different entities in the United States hold responsibility for implementing ESG disclosure regulations. The Securities and Exchange Commission (SEC) "is a principal regulator in the public market; it has taken the lead role in issuing ESG disclosures and related enforcement." Public traded companies can mandate the disclosure of material ESG risks and risks associated with climate change. It is keenly aware of investor demand for ESG information, and ESG regulations are counted among its priorities. In addition, SEC cooperates with ISSB; International Sustainability Standards Board; which works on setting standards like definitions, metrics, or methodologies for ESG rating providers. The Department of Labor (DOL) is responsible for providing guidelines and clarifying the use of ESG information in investment decisions. "It allowed plan fiduciaries to consider climate change and ESG factors when they select retirement investments and exercise shareholder rights, such as proxy voting." The Sustainability Accounting Standards Board (SASB) provides reporting CSR and ESG data guidelines. The Principles for Responsible Investment (PRI) also supports guidelines issued by the United Nations Global Compact (UNGC).

Greenwashing in the USA. Companies in the USA increasingly face pressure from various stakeholders, including customers, shareholders, and regulators, to improve their disclosure of their ESG goals and performance. This is driven by the growing importance of ESG considerations in investment decision-making and increased public awareness and concern about climate change and social justice.

Also, they are grappling with increasing legal actions by consumers and local governments. These litigations predominantly focus on the accusation of greenwashing, where a company's ESG assertions are purportedly inconsistent with the facts. Several states have established a legal framework regulating the scope of a company's ESG claims, imposing certain limitations on such proclamations.

11.5 Results and Implications

Several salient points can be inferred from the above comprehensive examination regarding ESG disclosure regulations and greenwashing practices in the USA and the EU.

11.5.1 EU Regulations Are More Advanced Than the USA

The EU and the USA have different approaches to ESG and greenwashing regulations, and it can be challenging to make a direct comparison. However, it is evident that the EU is more advanced than the USA regarding ESG disclosure and greenwashing regulations. The EU has taken a more comprehensive, specific, prudential, and conduct-based guidelines approach to ESG regulations, with a range of initiatives to promote sustainability and transparency in the financial sector. It has adopted a taxonomy of sustainable economic activities, which provides a framework for identifying and classifying environmentally sustainable investments. For example, in the EU, the materiality threshold for ESG disclosures is not associated with financial considerations. While in the United States, companies' ESG data are generally considered material and subject to compulsory disclosure only in evident financial consideration. The EU also requires companies to report on their ESG performance and has introduced regulations to combat greenwashing and ensure that companies are held accountable for their environmental claims.

In contrast, the USA has been slower to adopt comprehensive ESG regulations, with many initiatives taken by the private sector rather than government regulations. Nevertheless, there are indications that the situation is changing through significant regulatory developments in the US, noticeable by a surge of new ESG initiatives and proposals led by the SEC, the Biden Administration, and state-level regulations. In addition, establishing the Climate and ESG Task Force in 2021 addresses climate and ESG-related risks in the financial markets and the range of initiatives to promote sustainable investing, such as the Principles for Responsible Investment and the Sustainable Accounting Standards Board.

Both regions are making strides in this realm and indicating the increasing significance of sustainability and social responsibility in the financial sector. According to Bloomberg data analyses, "if there is a 15% growth, ESG assets under management could reach more than a third of the expected global assets under management of USD 140.5 trillion by 2025, totaling USD 53 trillion globally." The enlargement of ESG assets in the US during 2021 was significant, which might make the US the industry-leading light in the future, followed by Japan, despite Europe currently accounting for half of the global ESG assets.

It is challenging to forecast with certainty whether countries will adopt EU ESG regulations more than the USA, as it depends on diverse elements such as political, economic, and cultural considerations. However, the EU has been a leader in ESG regulation, and its regulations have been superior and comprehensive to those of the USA. This makes the EU regulations more attractive to countries pursuing to begin or enhance their own ESG regulations. Furthermore, the EU's emphasis on sustainable finance and its efforts to sponsor the global adoption of ESG standards may further raise the appeal of its regulations. The decision will depend on how countries view the effectiveness of the EU's regulatory framework. The EU's approach presents new concepts and potential solutions, often serving as a model for legislation in other countries seeking to reflect these ideas.

11.5.2 Mutual ESG Disclosure Challenges Between the EU and the USA

Even though the EU and the USA have taken different approaches to regulating ESG disclosure, they share similar obstacles in implementing and advancing ESG principles. This implies that the advanced regulations put in place by both regions have not fully solved the challenges related to ESG disclosure.

The main common challenge in ESG disclosure are: (1) The absence of universally accepted definitions and criteria. (2) Misconceptions about ESG integration. (3) The lack of a standardized, transparent reporting framework and unified data agency. (4) ESG data not capturing actual performance. (5) The absence of external data audit. (6) The lack of a global governing body.

11.5.3 Importance of International Standardization Regulations

Diverging regulations between markets will end in tensions between countries and companies. Differences in regulations may lead to regulatory confusion for companies that may not be fully integrated into the global economy or have less exposure to market regulations. Therefore, obtaining international standardized regulations on ESG is crucial as it can solve several difficulties and prevent greenwashing. As discussed above, the main challenges of the ESG topic are the lack of a universally accepted definition of ESG and ESG factors, a universal ESG reporting framework, and a mutual governing body. International standardized regulations ensure the unification of the mentioned above, presenting a solution for the challenge. As a result, it guarantees the obligation to implement ESG regulations in all member states. This can assist in blocking companies from exploiting loopholes in regulations or engaging in practices that damage the environment or society. It also ensures that companies are accountable for their environmental and social impact and transparent about their sustainability efforts. This can help prevent companies from engaging in greenwashing and increase investor confidence in the accuracy and reliability of ESG reporting. Finally, international ESG regulations, such as the United Nations' Sustainable Development Goals, can promote sustainable development and support achieving global sustainability objectives. It could generate the work to create a reliable and accessible ESG and climate data framework that fulfills the necessity to accelerate the implementation of a sustainable economy. Diverging regulations between markets will end in tensions between countries and companies.

On the other hand, the study acknowledges the existing criticism and different assessments of international standardized regulations for ESG and greenwashing. Some countries can consider legislating international regulations difficult and time-consuming, as it would require agreement and coordination between multiple countries. Different regulations could ease the implementation and enforcement of the regulations, as they would be modified to each country or region's specific needs and capacities. Detecting greenwashing cases could be more manageable as regulators, and investors would better understand and be more familiar with the local regulations. Furthermore, not all countries may have the financial resources or capabilities to adopt an international strategy for ESG and greenwashing regulations. One notable example is former U.S. President Donald Trump. During his time in office, his administration withdrew the USA from the Paris Climate Agreement and turned back various environmental regulations. Moreover, his administration countered specific international ESG standards with the excuse that they might hurt American businesses and limit economic growth.

Some companies may refuse such regulations for fear of increased costs and complex reporting requirements. Additionally, some critics can argue that standardized ESG regulations could lead to a "box-ticking" approach to sustainability, where companies focus more on meeting regulatory requirements than implementing meaningful sustainability initiatives (Khoffash and Awwad 2024; Awwad 2025a, b).

This could undermine the true purpose of ESG reporting, which is to encourage companies to adopt more sustainable and ethical practices.

11.6 Limitations

The paper has certain limitations that must be acknowledged. Firstly, the definition of greenwashing employed is restricted to the definition put forth by ESMA, which is just one of many widely accepted definitions of greenwashing. Secondly, the study is limited in scope as it focuses solely on the phenomenon in relation to ESG disclosure regulations, while greenwashing can be studied under different circumstances or in relation to other topics. Thirdly, the comparative study is limited to exploring the regions of the EU and the USA. However, it is crucial to recognize that developing countries play a pivotal role in global sustainability efforts, and many of them are heavily impacted by greenwashing and have a greater need for ESG regulations. Fourthly, the paper's examination of literature and regulation is conducted under a specific time frame, and studies conducted under different time frames may produce varying results.

Despite these limitations, the importance of developing international standardized regulations for ESG and greenwashing cannot be understated. The increasing number of international initiatives focused on ESG regulations demonstrates the urgency and importance of addressing these issues.

11.7 Recommendation for Future Research

Given the novelty of the topic, the paper focuses on a specific definition of greenwashing along with ESG disclosure regulations in the USA and European Union. While studying the USA and the EU approach is valuable, a comparative study of ESG disclosure regulations across different countries and regions could provide further insights. It is essential to consider the perspectives and experiences of developing countries in sustainability discussions, given their weak regulatory policies and greater vulnerability to greenwashing. Moreover, greenwashing comes under distinct definitions and types, which warrant different examinations under different examples.

Furthermore, the legislation and regulations regarding ESG and greenwashing are relatively recent and will be enforced in the coming months or years. Therefore, measuring the effectiveness of current ESG and greenwashing regulations is vital, and future research could explore their impact and identify areas for improvement. Such research could inform the development of international standardized regulations for ESG and greenwashing, essential for promoting transparency, accountability, and sustainability in business practices.

References

Awwad BS (2022) The role of environmental disclosure in enhancing the market value of industrial and service companies listed on the Palestinian stock exchange. Int J Entrepr 26(2):1–12. ISSN: 1939-4675

Awwad BSAL (2024) Governance with relationship between entrepreneurship and economic growth in Palestine. Int J Law Manag 66(2):259–287

Awwad BS (ed) (2025) Sustainability in light of governance and artificial intelligence applications. Emerald Group Publishing.

Awwad B (2025) Climate finance and sustainability: a story told through the Journal of Cleaner Production. EuroMid J Bus Tech-Innov 31–47.

Awwad B, El Khoury R (2024) Information technology governance and bank performance: evidence from Palestine. J Decis Syst 33(2):311–334

Awwad BS, Alkababji MW, Zidan S (2023a) Governance and social responsibility: Palestinian Investment Fund. Bus Econ Res 13(4):185–215

Awwad BS, Binsaddig R, Kanan M, Al Shirawi T (2023b) Women on boards: an empirical study on the effects on financial performance and corporate social responsibility. Competitiv Rev: Int Bus J 33(1):147–160

Awwad BS, Razia BS, Razia AS (2024) Digital transformation under the governance of Palestinian banks. Discover Sustain 5(1):76

Boubaker S, Liu Z, Zhan Y (2022) Customer relationships, corporate social responsibility, and stock price reaction: lessons from China during health crisis times. Financ Res Lett 47:102699. https://doi.org/10.1016/j.frl.2022.102699

Berrone P, Fosfuri A, Gelabert L, Gomez-Mejia LR (2015) Necessity as the mother of "green" inventions: institutional pressures and environmental innovations. Strategic Manage J 36(6):891–909. https://doi.org/10.1002/smj.2241

Bruton GD, Ahlstrom D, Li HL (2010) Institutional theory and entrepreneurship: where are we now and where do we need to move in the future? Entrep Theory Pract 34(3):421–440. https://doi.org/10.1111/j.1540-6520.2010.00390.x

Cho CH, Laine M, Roberts RW, Rodrigue M (2015) Organized hypocrisy, organizational façades, and sustainability reporting. Acc Organ Soc 40:78–94. https://doi.org/10.1016/j.aos.2014.12.003

Clementino E, Perkins R (2020) How do companies respond to environmental, social and governance (ESG) ratings? Evidence from Italy. J Bus Ethics 171(2):379–397. https://doi.org/10.1007/s10551-020-04441-4

Crilly D, Hansen M, Zollo M (2016) The grammar of decoupling: a cognitive-linguistic perspective on firms' sustainability claims and stakeholders' interpretation. Acad Manag J 59(3):705–729. https://doi.org/10.5465/amj.2015.0171

de Freitas Netto SV, Sobral MFF, Ribeiro ARB, Soares GRDL (2020) Concepts and forms of greenwashing: a systematic review. Environmental Sciences Europe 32:19. https://doi.org/10.1186/s12302-020-0300-3

Delmas MA, Burbano VC (2011) The drivers of greenwashing. Calif Manag Rev 54(1):64–87. https://doi.org/10.1525/cmr.2011.54.1.64

Doan H, Sassen R (2020) The relationship between environmental performance and environmental disclosure: a meta-analysis. J Ind Ecol 24(5):1231–1242. https://doi.org/10.1111/jiec.13002

Eccles RG, Klimenko S (2019) The investor revolution. Harvard Bus Rev. https://hbr.org/2019/05/the-investor-revolution

European Securities and Markets Authority (ESMA) (2021) Final report: guidelines on disclosure of ESG risks and opportunities by issuers. https://www.esma.europa.eu/sites/default/files/library/esma32-32-1389_final_report_on_esg_disclosure_guidelines.pdf

Flammer C (2014) Does product market competition foster corporate social responsibility? Evidence from trade liberalization. Strateg Manag J 36(10):1469–1485. https://doi.org/10.1002/smj.2307

Global Sustainable Investment Alliance (GSIA) (2020) Global sustainable investment review 2020. http://www.gsi-alliance.org/trends-report-2020/

Haque F, Ntim CG (2018) Environmental policy, sustainable development, governance mechanisms and environmental performance. Bus Strateg Environ 27(3):415–435. https://doi.org/10.1002/bse.2007

Hartzmark SM, Sussman AB (2019) Do investors value sustainability? A natural experiment examining ranking and fund flows. J Finance 74(6):2789–2837. https://doi.org/10.1111/jofi.12841

Khoffash S, Awwad BS (2024) The challenges and opportunities of green financing in Palestine from the perspective of banks. J Infrastruct Policy Dev 08(02):1–20. https://doi.org/10.24294/jipd.v8i12.8800

La Torre M, Huse C, Salvato C (2020) Corporate sustainability, governance, and shareholder activism in Europe and the United States. J Manag Stud 57(7):1384–1410. https://doi.org/10.1111/joms.12563

Lyon TP, Maxwell JW (2011) Greenwash: Corporate environmental disclosure under threat of audit. J Econ Manag Strategy 20(1):3–41. https://www.researchgate.net/publication/5134228

Lyon TP, Montgomery AW (2015) The means and end of greenwash. Organ Environ 28(2):223–249. https://doi.org/10.1177/1086026615575332

Mateo-Márquez AJ, De Fuentes C, Carrasco RA, Rodríguez-Sánchez I (2022) An international empirical study of greenwashing and voluntary carbon disclosure. J Clean Prod 363:132567. https://doi.org/10.1016/j.jclepro.2022.132567

Miralles-Quirós MM, Miralles-Quirós JL, Redondo Hernández J (2019) ESG performance and shareholder value creation in the banking industry: International differences. Sustainability 11(5):1404. https://www.mdpi.com/2071-1050/11/5/1404

Ortas E, Álvarez I, Jaussaud J, Garayar A (2015) The impact of institutional and social context on corporate environmental, social and governance performance of companies committed to voluntary corporate social responsibility initiatives. J Clean Prod 108:673–684. https://doi.org/10.1016/j.jclepro.2015.05.133

Pope S, Wæraas A (2015) CSR-washing is rare: a conceptual framework, literature review, and critique. J Bus Ethics 137:173–193. https://doi.org/10.1007/s10551-015-2546-z

Roulet TJ, Touboul S (2015) The intentions with which the road is paved: attitudes to liberalism as determinants of greenwashing. J Bus Ethics 128:305–320. https://doi.org/10.1007/s10551-014-2097-8

Steiner G, Geissler B, Schreder G, Zenk L (2018) Living sustainability, or merely pretending? From explicit self-report measures to implicit cognition. Sustain Sci 13:1001–1015. https://doi.org/10.1007/s11625-018-0561-6

Sullivan R, Mackenzie C (2020) Standardizing ESG reporting: an urgent global need. J Bus Ethics 161:9–24. https://doi.org/10.1007/s10551-018-3943-1

Tateishi E (2017) Craving gains and claiming "green" by cutting greens? An exploratory analysis of greenfield housing developments in Iskandar Malaysia. J Urban Aff 40(1):1–24. https://doi.org/10.1080/07352166.2017.1355667

Townshend T (2014) Climate change legislation: China has politics, too. Nat Clim Chang 4:931. https://doi.org/10.1038/nclimate2415

Uyar A, Karaman AS, Kuzey C (2020) Is corporate social responsibility reporting a tool of signaling or greenwashing? Evidence from the logistics industry. J Clean Product 253:119888

Yu EP-Y, Luu BV, Chen CH (2020) Greenwashing in environmental, social and governance disclosures. Res Int Bus Financ 52:101192. https://doi.org/10.1016/j.ribaf.2020.101192

Zhang L, Liu Y, Liu Y, Wang Y (2023) The influence of greenwashing perception on green purchasing intentions: the mediating role of green word-of-mouth and moderating role of green concern. J Clean Prod 187:740. https://doi.org/10.1016/j.jclepro.2022.132567

Part III

Entrepreneurship, Education, and Social Development

12 Islamic Finance at a Crossroad: Development, Inclusive Growth, and the Governance Challenge

Majdoulene Kerish

Abstract

Dual-purpose financial systems that integrate Islamic and conventional finance offer an untapped potential for inclusive and sustainable growth. While conventional banking often holds the majority of assets and plays a significant role, the distinct features of Islamic finance, such as its emphasis on risk-sharing and financial inclusion, can complement the shortcomings of the conventional system. However, the realization of this potential hinges on coherent governance arrangements, legal harmonization, and institutional capacity, which are often underdeveloped even in pioneer nations. This paper proposes a conceptual framework to assess the dual systems and investigates whether they can contribute to development and inclusive growth. Through a comparative case study analysis of Malaysia, a state-led model with strong institutional frameworks, and Indonesia, a more gradual, market-driven approach facing ongoing challenges, we identify the governance conditions that are prerequisites for success. The paper concludes that a successful dual-purpose system requires deliberate cultivation of a stable legal and regulatory environment to ensure philosophical alignment and economic performance.

Keywords

Islamic finance • Financial duality • Inclusive growth • Development • Governance

M. Kerish (✉)
Loughborough University, Loughborough, United Kingdom
e-mail: majk3@cantab.ac.uk

University of Cambridge, Cambridge, United Kingdom

B. Awwad and R. A. Awwad (eds.), *Technology and Innovation Towards a Future of Business Sustainability*, Contributions to Environmental Sciences & Innovative Business Technology, https://doi.org/10.1007/978-981-95-6469-9_12

12.1 Introduction

The remarkable and sustained growth of Islamic finance with assets reaching $5.4 trillion by 2024 and an annual growth rate of 10%, has captured the attention of both policymakers and researchers (Standard Chartered). Proponents of Islamic finance highlight its potential to foster a more sustainable system. They point to its reliance on Shariah-compliant instruments with emphasis on equitable risk-sharing, social justice, and a direct link to real economic activity. Studies on countries like India and Indonesia have shown that Islamic finance can serve vulnerable populations, including farmers and SMEs, thereby supporting inclusive economic growth (Saida and Frikha 2016; Ghanavati et al. 2024).

However, despite this growing body of research, a significant gap remains. The existing literature often treats Islamic finance in isolation or frames it as a complete alternative to conventional systems. This overlooks the common reality in developing economies: the coexistence of both systems in a dual-purpose financial system. The question is not whether Islamic finance can succeed on its own, but whether it can thrive alongside a conventional system to create a more resilient and inclusive economy and contribute to development. Critically, there are also counterarguments regarding the economic consequences of Shariah-compliant restrictions (Kuran 1995; Awad et al. 2025), which necessitate a more nuanced understanding of the conditions for success.

This paper addresses this gap by investigating the research question: Can dual-purpose financial systems contribute to development and improve inclusive growth outcomes in developing economies, and what governance conditions are required for success? We argue that the full potential of such systems is contingent upon coherent governance, legal harmonization, and institutional readiness. The objective is to move beyond a binary view of financial systems and propose a framework that provides policymakers with insights on how to harness the benefits of duality while navigating its complexities. The paper will first outline a conceptual framework, then apply it to comparative case studies of Malaysia, and Indonesia to identify the lessons learnt.

12.2 Methodology

This paper uses a multi-faceted methodology to address its central research question. It combines a comparative case study analysis with analysis of existing literature and the development of a conceptual framework.

The core of the research is a **comparative case study analysis** of two models that enable an examination of governance, legal and institutions:

- **Malaysia**: A country with a government-led dual system.
- **Indonesia**: A country with a gradual, market-driven approach to duality.

This approach allows for the identification of critical governance variables that distinguish successful dual systems from those facing structural constraints. The analysis is informed by a synthesis of literature drawn from several key areas: Islamic finance and growth studies, zakat governance models, and public finance governance frameworks. This approach ensures that the analysis is grounded in the unique principles of Islamic finance and the broader principles of sound public administration.

The conceptual framework is developed to systematically evaluate the case studies. This framework combines four key dimensions: financial system pluralism, legal harmonization, governance infrastructure, and philosophical coherence. This allows for a structured assessment of the enablers and bottlenecks of each country's approach. The research draws on a variety of sources, including peer-reviewed academic literature, international finance standards from bodies such as the Islamic Financial Services Board (IFSB) and the Accounting and Auditing Organization for Islamic Financial Institutions (AAOIFI), and secondary data on macroeconomic indicators and social finance outcomes.

12.3 Conceptual Framework: The Pillars of a Dual-Purpose System

The success of a dual-purpose financial system hinges on the development of four key dimensions. The framework suggests that the synergy between these dimensions determines whether the system becomes a catalyst for inclusive growth or a source of institutional fragility (Fig. 12.1).

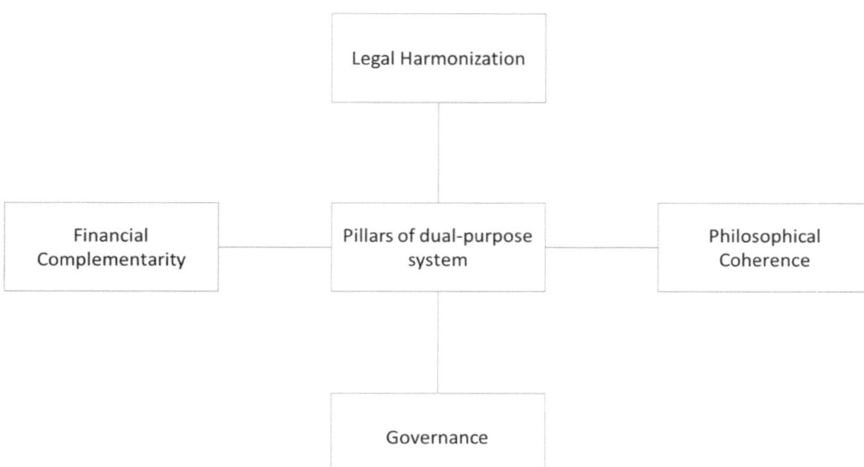

Fig. 12.1 Conceptual framework

12.3.1 Financial Complementarity

This dimension assesses how Islamic finance features complement the conventional system. Rather than viewing Islamic finance as a mere alternative, we examine its unique contributions: risk-sharing models (Mudarabah, Musharakah); finance based on Islamic principles which screens out what it perceives to be harmful or non-productive activities (Riba: which means interest; Gharar which means excessive uncertainity and ambiguity; Maysir which means gambling); and financial inclusion, which targets underserved populations through profit-and-loss sharing or microfinance schemes.

12.3.2 Legal Harmonization

The coexistence of two financial systems requires a legal framework that merges Shariah-compliant contracts with commercial and civil law. This involves creating clear rules for contract enforcement, establishing a predictable dispute resolution mechanism, and ensuring a level playing field for both Islamic and conventional institutions.

12.3.3 Governance Infrastructure

This includes the institutional capacity of central banks and financial regulators to supervise and regulate both conventional and Islamic institutions. It also requires the establishment of independent Shariah supervisory boards to ensure product compliance, as well as specialized dispute resolution bodies.

12.3.4 Philosophical Coherence

Beyond the technical and legal aspects, a successful dual system must embody a deeper philosophical coherence. This means aligning the financial system with the core principles of Islam. The author of this paper argues that the most critical principles for the success of a dual system-especially as Islamic finance products merge with and evolve within the market, are those underpinning the institution of zakat: justice (‘adl), mercy (rahma), solidarity (takaful), and the recognition of the poor’s legal rights (huquq).

These four elements are chosen specifically because they capture the unique dual character of *zakat* as both an act of worship (*ibadah*) and a social transaction (*mu’amalah*). While modern economic systems offer instruments like taxation and charity, none fully replicate this integrated, rights-based approach. Secular models are often efficiency-driven or voluntary, whereas the philosophy of *zakat* mandates a structural redistribution based on *mercy* and the recognition of the poor’s *rightful* claim. This distinction is crucial, as it provides a foundation for a financial system that prioritizes social legitimacy and inclusive growth alongside economic performance. By grounding the conceptual framework in these principles, this paper moves beyond a purely technical analysis to explore whether the religious essence of Islamic finance can be preserved and effectively integrated into modern states for development.

12.4 Findings and Discussion

12.4.1 Malaysia: A Model of Duality

Malaysia's experience provides a compelling case study of how a strong legal and governance infrastructure can enable a successful dual-system for development. The country's Islamic finance sector is ranked third globally in the world with $521 billion in assets, behind only Saudi Arabia and Iran, which both have significantly larger Muslim populations. Furthermore, Malaysia dominates the global *Sukuk* (Islamic bond) market, holding 46.6% of the world's total assets, and accounts for 20% of the total global Islamic finance sector dealings (World Bank 2020). The true value of Malaysia as a model lies in how its government-led strategy built the legal and institutional frameworks to operationalize its philosophical principles.

Islamic Finance in Malaysia can be traced back to 1963 with the introduction of Tabung Haji, an entity to help Muslims save for Hajj and the establishment of the first Islamic Bank in 1983 with Bank Islam Malaysia Berhad (Balarabe et al. 2023). These events were foundational steps in long-term growth and a diversified financial system. The result is a dual system, with 45.6% Islamic finance out of the overall banking system (DDCAP 2024), that has contributed to positive macroeconomic impacts and financial resilience.

Islamic Finance contributed to Malaysia's growth when compared to the Tiger Cub Economies of Southeast Asia (Indonesia, the Philippines, Thailand, and Vietnam), and didn't hinder it. The true test of the financial system came with the Asian economic crisis of 1997. While all the Tiger Cubs were hit hard, Malaysia's dual-purpose system showed a degree of resilience (Rethink 2023). While Vietnam's average GDP growth rate was higher than Malaysia's post-crisis average (1999–2005), at 6.58% and 5.54%, respectively, this observation doesn't negate the success of Malaysia's dual-purpose financial system. Instead, it shifts the focus of the argument from growth velocity to stability and resilience (Fig. 12.2).

The dual-purpose system was not designed to achieve the highest possible growth rates but rather to create a more diversified economy. While Malaysia's economy contracted in 1998, its recovery in the years that followed (with a growth rate of 8.7% in 2000) demonstrated a fundamental strength that allowed it to weather a global slowdown. Even with the slower growth rate in 2001, compared to the Tiger Cub, Islamic finance continued to grow, providing a stable alternative source of financing and liquidity that helped prevent a deeper recession. (Ahmad Mokhtar et al. 2006) (Fig. 12.3).

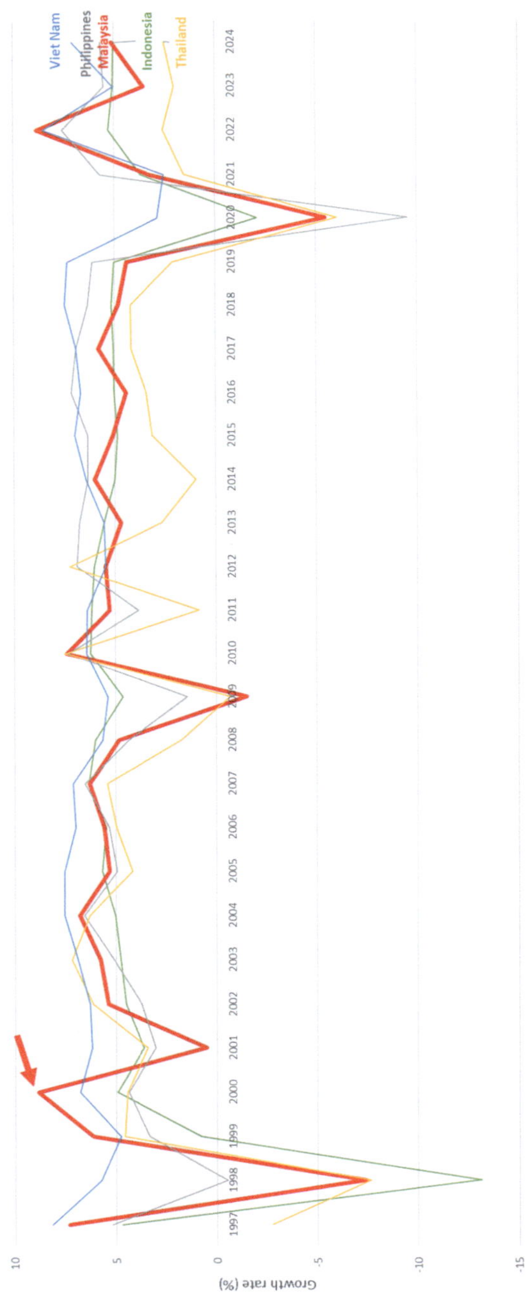

Fig. 12.2 GDP growth rates, 1997–2024. *Source* World Bank Data

	As at end of						
	1997	1998	1999	2000	2001	2002	2003
Total Assets	**17,881.3**	**21,183.1**	**33,558.7**	**42,725.3**	**55,605.4**	**63,321.9**	**77,390.6**
Full-fledged Islamic Bank	5,202.1	5,698.4	11,724.2	14,008.9	17,404.8	20,159.6	20,929.7
Islamic Windows:							
Commercial Bank	9,078.0	11,385.2	15,589.1	20,058.5	27,026.1	29,109.8	36,830.0
Finance Company	2,924.4	3,321.4	4,806.1	7,149.9	9,821.6	12,622.9	17,915.1
Merchant Bank	676.8	778.1	1,439.3	1,508.0	1,352.9	1,429.6	1,715.8
Total Deposits	**9,895.2**	**15,172.1**	**23,695.7**	**33,650.7**	**44,743.8**	**49,553.9**	**55,919.7**
Full-fledged Islamic Bank	3,223.4	4,039.7	9,685.2	11,301.6	14,375.6	16,421.2	17,583.7
Islamic Windows:							
Commercial Bank	5,153.2	8,415.2	10,576.0	16,089.4	22,031.0	23,353.9	26,518.7
Finance Company	1,170.2	2,110.7	3,033.1	5,392.6	7,663.7	9,094.6	10,965.6
Merchant Bank	348.4	606.5	401.4	867.1	673.5	684.2	851.7
Total Financing	**10,749.4**	**10,461.1**	**13,723.7**	**20,816.1**	**28,317.6**	**36,717.7**	**48,615.4**
Full-fledged Islamic Bank	3,350.7	3,471.4	5,029.5	6,423.4	7,671.0	9,158.2	9,764.5
Islamic Windows:							
Commercial Bank	4,705.8	4,702.8	4,920.5	8,533.6	12,257.6	16,706.4	22,324.3
Finance Company	2,189.9	1,878.4	2,995.5	5,089.8	7,617.4	10,049.6	15,745.8
Merchant Bank	503.0	408.5	778.2	769.3	771.6	803.5	780.8

Fig. 12.3 Total assets, deposits and financing of Islamic banking (RM million)

Malaysia's success stems from its state capacity and its commitment to both financial sector development and philosophical alignment, as we see below when looking at the components of the conceptual framework.

Financial Complementarity

Malaysia has seen an increase in financial inclusion since the introduction of a dual system, particularly for the population who were previously reluctant to engage with conventional finance for religious reasons. Data from 2018 shows that 85% of adult Malaysians were financially included, with Islamic finance playing a key role. However, the Malaysian system has its challenges. The data suggests that while Islamic finance has been a major contributor to financial inclusion, it could still go further. A small portion of the population (1.6%) still cites religious concerns as a barrier to financial engagement, suggesting a lack of awareness about the availability of Shariah-compliant products or the differences between conventional and Islamic financial services (World Bank 2020). Furthermore, while the system offers a range of Shariah-compliant products, a significant challenge remains in balancing commercial viability with the philosophical tenets of Islamic finance. A 2023 report shows that financing based on debt-like instruments accounted for over 80% of total Islamic banking assets, whereas profit and loss sharing models remain a small percentage (INCEIF 2023). The reliance on debt-based contracts suggests that the system's contribution is more about Shariah compliance than fundamentally altering the risk profile or redistributing the capital in a truly participatory way.

Legal Harmonisation and Governance

Malaysia's legal framework, anchored by the Islamic Financial Services Act (IFSA) 2013, serves as a foundational enabler for its successful dual-purpose system. The Act provides a clear, legally binding framework that ensures Islamic finance operates alongside conventional law without conflict, minimizing ambiguity and building trust. This legal framework is complemented by a governance infrastructure that ensures the principles enshrined in the law are applied consistently and credibly across the financial system.

- Elevating Shariah from principles to Law and legally binding: Section 28 of the IFSA 2013 is a pivotal provision, as it legally obligates institutions to ensure compliance with Shariah. This transforms Shariah from a set of guidelines into a mandatory legal requirement, providing a strong legal foundation that is crucial for building a credible and stable financial system. This clarity is a key enabler of inclusive growth, as it builds confidence among consumers, including those who were previously hesitant to engage with a system perceived as lacking legal certainty.
- Centralized Interpretation and Consistency: The establishment of the Shariah Advisory Council (SAC), as referenced in Section 29, ensures a top-down governance structure where the SAC's rulings are binding. This centralized authority prevents the fragmentation and inconsistent interpretations of Shariah that can plague dual systems, thereby providing a consistent and certain regulatory environment for both financial institutions and their customers.
- Robust Governance Infrastructure: The Act mandates the creation of internal Shariah committees (Section 30) within each institution and requires regular Shariah compliance audits (Sections 37–38). This institutional framework ensures that Shariah principles are embedded at all levels of an organization and are subject to regular, independent oversight. This robust internal governance, combined with the Central Bank's regulatory power, acts as a powerful check against institutional capture and under-regulation, which are common governance challenges in developing economies.
- Avenues for Dispute Resolution: The Act's provisions for dispute resolution are critical for consumer protection and legal certainty. While the Financial Ombudsman Scheme (Part IX, Section 138) provides an accessible and fair mechanism for resolving consumer complaints, the inclusion of the High Court and the provisions for civil and criminal actions (Part XVI) provide multiple avenues for legal recourse. This layered approach to dispute resolution further strengthens trust in the system and demonstrates a commitment to upholding the rule of law, which is a hallmark of good governance.

Indeed, the IFSA 2013 was not a revolutionary catalyst for sudden, dramatic growth, but an evolution that brought about crucial stability (Fig. 12.2). While the Act did not cause a sharp or surprising surge in growth, it changed the quality of that growth as shown in the statistical analysis of the growth rate. The data reveals a sharp decrease in volatility post-implementation: the standard deviation plummeted from 2.733 in the 2000–2012 period to just 0.67 in the 2013–2015 period.[1] This significant reduction in volatility demonstrates that the Act's comprehensive legal and governance framework contributed to de-risking the sector.

In 2020, Malaysia's GDP growth rate was − 5%, a sharper contraction than its fellow dual-system economy, Indonesia, which saw a contraction of − 2%. While this disparity might suggest a weakness in the Malaysian model, it instead highlights a fundamental difference in economic structure. Malaysia's economy is highly dependent on exports, which constituted over 60% of its GDP pre-COVID (World Bank Data). The global supply chain disruptions and plummeting international demand in 2020 were a direct and unavoidable blow to this economic model. In contrast, Indonesia's more domestically driven economy was comparatively insulated. This shows that while the economic model made Malaysia more vulnerable to a global trade shock, the dual-purpose financial system contributed to shock responsiveness and recovery.

This legal infrastructure is the very mechanism that allows the philosophical principles to be operationalised, as we see in the next section.

Philosophical Coherence

While the dual system in Malaysia is not without its challenges, particularly in the realm of financial complementarity and risk-sharing (as we saw in an earlier point in the case study), its approach to public finance through zakat offers a compelling case of philosophical coherence. The Malaysian model is distinguished by its ability to operationalize zakat as a public finance tool while safeguarding its ethical foundations of justice, mercy, and solidarity.

The principle of Ḥaqq (Legal Right) is a foundational element of Malaysia's system. Zakat is not merely a voluntary religious act but a legally enforceable right, enshrined through state-level Islamic Religious Council Enactments. For instance, the Administration of the Religion of Islam (State of Selangor) Enactment 2003 mandates zakat payment and outlines the powers of the Lembaga Zakat Selangor (Zakat Board of Selangor) to collect and distribute funds (Selangor 2003). Similar laws exist in other states like the Zakat Enactment (State of Kedah) 2010 (Haji-Othman et al. 2021). This framework ensures that the poor have a legal claim to zakat, transforming it into a structured pillar of a welfare state rather than a

[1] The decision to focus on the 2013–2015 period for this analysis is deliberate. By isolating the period immediately following the Act's passage, we can more confidently attribute the significant decrease in volatility to the implementation of the new legal and regulatory framework. A longer time frame after 2013 would introduce new confounding variables and economic events that could distort the analysis, making it more difficult to demonstrate a clear cause-and-effect relationship.

discretionary act of charity. They act as the legal mechanism that operationalizes the philosophical principle of Ḥaqq.

This legal basis empowers a robust governance structure, which in turn reinforces Justice (*ʿAdl*), Mercy (*Raḥmah*), and Solidarity (*Takaful*). The Malaysian system is characterized by structured collection and community-based distribution mechanisms. For example, institutions like the Kedah State Zakat Board have established systematic processes for both collecting zakat and delivering it to beneficiaries through local, community-based programs. This model ensures targeted distribution, which aligns with the principle of justice.

In a comparative context, the growth and institutional maturity of the takaful industry in Malaysia serves as a indicator of the system's survival in today's states. The country's takaful sector has consistently demonstrated strong performance, with general takaful seeing a 10.5% year-on-year contribution growth in the first half of 2024, outpacing the growth of the conventional non-life insurance sector. While the family takaful market share saw a slight decline in 2024, Malaysia was previously recognized as the world's largest family takaful market in 2015, highlighting a strong and established foundation (Insurance Asia 2025). This institutionalized and community-integrated approach demonstrates that sound governance and institutional maturity are technical prerequisites, and essential for preserving the philosophical essence of Islamic social finance and enabling its contribution to inclusive growth.

12.4.2 Indonesia: A Gradual but Challenged Approach

Figure 12.2 might have sparked some questions on the situation in Indonesia, which is also a country with a dual-purpose system. Hence, this chapter will analyse how Islamic finance contributes to development in Indonesia.

Indonesia's approach to Islamic Finance has been more gradual and decentralized, often driven by community initiatives (Faizi 2023). Islamic finance is 10.95% of the total market share and assets reached USD 4.5 billion with 9% growth rate in 2023 (OJK 2023).

The country's history is marked by a pivotal event: the 1998 Asian Financial Crisis. At that time, Islamic finance was in its infancy, with the first Islamic bank, Bank Muamalat, established only in 1992. This stage meant that the absence of a developed parallel system made Islamic finance a non-factor in mitigating the crisis, and the entire conventional financial sector suffered the collapse with a − 13% growth rate in 1998 (Fig. 12.2).

Financial Complementarity

Indonesia's Islamic finance sector demonstrates a dual nature: a successful social finance segment and a struggling commercial banking sector in terms of inclusivity. The social finance system, which includes institutions like Badan Amil Zakat Nasional (BAZNAS), actively works to alleviate poverty and empower microenterprises. It utilizes instruments such as zakat and qard al-hasan (benevolent loans) to drive sustainable development at the grassroots level. This success in the social sector, however, contrasts with the broader Islamic banking sector's limited reach.

The Islamic financial inclusion index for Indonesia is extremely low, with a figure of 0.18%. This index, which measures the accessibility, availability, and usage of Islamic banking services, points to significant limitations in the sector's reach, service presence, and actual use by the population (Hamidi et al. 2025). This low inclusion is particularly pronounced in poorer regions, revealing a strong link between financial access and human development. For example, a paper noted that: East Nusa Tenggara, one of the country's poorer regions, has the lowest average Islamic financial inclusion index, at just 0.002, while DKI Jakarta, a province with a high Human Development Index, has a significantly higher average inclusion rate of 0.577 for the years 2014–2017. These regional disparities show that the Islamic finance sector's benefits have not been equitably distributed, with a clear need to expand its reach and improve access for low-income populations in underdeveloped areas (Ali et al. 2019). Moreover, the sector's heavy reliance on debt-based products over risk-sharing models, similar to Malaysia, also suggests a gap between its philosophical ideals and its reality.

Legal Harmonization and Governance

The foundational Law No. 21/2008 on Sharia Banking was a critical step in establishing a formal, parallel system. However, its implementation highlights the challenges of a more decentralized, market-driven approach (Republic of Indonesia 2008).

The law's core strength lies in its definitions and regulatory mandates. It provides clear definitions for key Shariah-compliant contracts, or akad, such as wadi'ah (deposits), mudharabah (investments), and murabahah (sales and purchase), thereby giving the sector legal certainty. The law also establishes a clear governance structure, mandating the formation of a Sharia Supervisory Board (SSB) within each institution to ensure compliance with Shariah principles. This is a crucial first step toward formalizing religious oversight.

However, the Indonesian model faces a key challenge in fragmentation. Unlike Malaysia's Shariah Advisory Council (SAC) with its legally binding rulings on all financial institutions, Indonesia's system relies on fatwa from the Indonesian Ulama Council (MUI), which are then incorporated into regulations by Bank Indonesia. This tiered approach, while centralized in theory, can lead to inconsistencies and a lack of unified legal precedence. This legal dualism can create ambiguity for financial institutions and consumers, which, in turn, hinders the seamless integration of both financial models and limits the industry's full potential.

Furthermore, the law's provisions on dispute resolution highlight this fragmented approach. While it states that disputes are to be heard in a Religious Court, it also includes a provision allowing for alternative dispute resolution methods if parties have already agreed to them in their contract. This dual-path system, while offering flexibility, contrasts with Malaysia's more integrated civil court system with specialized Shariah-trained judges. The Indonesian model's reliance on contractual agreements for dispute resolution could lead to a less harmonized outcome, thereby underscoring the point that a strong legal framework alone is not enough; it requires a unified governance infrastructure to truly drive inclusive growth.

Philosophical Coherence

The principles of rights, mercy and justice are not abstract concepts in Indonesia but are legally mandated components of state-led social policy. Undang-Undang Nomor 23 Tahun 2011 (Law No. 23 of 2011) on Zakat Management institutionalizes the principle of huquq by legally establishing zakat as a right for the poor (mustahik), rather than a discretionary charitable act. This law created the National Amil Zakat Agency (BAZNAS), a government body tasked with the organized collection and distribution of zakat. This formal, state-backed structure effectively transforms the religious obligation into a rights-based system of redistribution. Furthermore, institutions like BAZNAS operationalize mercy and justice through "productive zakat" programs. Rather than merely providing unconditional cash transfers, these initiatives focus on economic empowerment, such as offering vocational training, micro-financing, and business development support. This dual focus demonstrates mercy by alleviating immediate hardship while simultaneously enacting justice by providing the tools for long-term self-sufficiency. Indonesia's zakat system thus stands as a successful example of how a state can use legal and institutional infrastructure to codify and enforce a deeply philosophical framework.

In contrast, the principle of takaful (solidarity) reveals the limitations of this model when applied to the commercial sector. While the philosophical concept of mutual risk-sharing is popular, its commercial manifestation-the Islamic insurance industry-remains marginal. Accounting for only around 10% of the total insurance market in 2024 (FitchRating 2025). The philosophical coherence of takaful as a cooperative model of social protection has not been fully translated into a compelling and competitive market offering. This disconnect suggests that even a deeply-held principle struggles to take root without a dedicated legal and governance framework that specifically nurtures commercial imperatives with the principles and objectives.

12.4.3 Cross-Cutting Themes

The comparative analysis of Malaysia and Indonesia reveals several key themes that are essential for understanding the success and challenges of dual-purpose financial systems. The Malaysian model demonstrates that a successful system is a function of strong state capacity. Its government-led strategy, anchored by legislation like the IFSA 2013, created the necessary legal and institutional infrastructure to operationalize Islamic finance and contribute to stability, as evidenced by the reduction in market volatility post-2013.

In contrast, the Indonesian case highlights the risks of fragmentation and a lack of unified governance. While Indonesia has been successful in its social finance segment, its market-driven, decentralized approach has led to a fragmented legal and regulatory landscape. This is visible in the inconsistent interpretations of Shariah and the dual-path system for dispute resolution, which has hindered the industry's full potential for commercial inclusivity and limited the growth of its takaful sector.

Both case studies also show the vital role of international standards from bodies like the Islamic Financial Services Board (IFSB) and the Accounting and Auditing Organization for Islamic Financial Institutions (AAOIFI) in providing a harmonized framework. However, these standards are most effective when supported by a strong domestic governance infrastructure capable of enforcing them consistently. Ultimately, the success of a dual system rests on its ability to create a synergy between institutional maturity and philosophical principles.

12.5 Conclusion

This paper set out to answer the research question: Can dual-purpose financial systems contribute to development and improve inclusive growth outcomes in developing economies, and what governance conditions are required for success? Based on our conceptual framework and the comparative analysis of Malaysia and Indonesia, the answer is a qualified **yes**. Dual-purpose financial systems have the potential to be a powerful catalyst for development, but their success is not inherent; it is contingent upon a deliberate governance and legal system.

Our findings lead to several key conclusions and policy recommendations:

- **Duality is more viable than rushed Islamization**. The success of the Malaysian model, with its long-term, government-led strategy, shows that a gradual and institutionally robust approach is superior to a rushed implementation. A system's ability to foster inclusive growth is directly tied to its institutional maturity and capacity to support both conventional and Islamic finance.
- **Invest in Regulatory Capacity and Shariah Governance**. The Malaysian experience with the IFSA 2013 and its centralized Shariah Advisory Council (SAC) demonstrates the importance of a clear and legally binding regulatory framework. Policymakers must invest in supervisory tools for Islamic finance and create unified Shariah governance structures to ensure consistency, build consumer confidence, and provide a stable environment for both financial institutions and customers.
- **Develop Hybrid Legal Mechanisms**. Legal harmonization is a prerequisite for success. As shown by the contrast between Malaysia's integrated civil court system with specialized judges and Indonesia's fragmented approach, effective dispute resolution mechanisms are crucial for legal certainty. Developing hybrid legal frameworks that provide clear avenues for contract enforcement is essential for reducing ambiguity and encouraging wider adoption of Islamic financial products.

- **Tailor reforms to country readiness**. The differing trajectories of Malaysia and Indonesia underscore that there is no one-size-fits-all model. Countries seeking to implement or expand a dual system must first assess their own institutional readiness, legal frameworks, and market dynamics rather than simply attempting to replicate another country's model.

References

Ahmad Mokhtar HS, Abdusllah N, Al Habashi SM (2006) Efficiency of Islamic banking in Malaysia: a stochastic frontier approach. J Econ Cooper 37–70

Ali M, Sakti MRP, Devi A (2019) Developing Islamic financial inclusion index for Islamic banks in Indonesia: a cross-province analysis. J Islamic Monet Econ Finance 5:691–712. https://doi.org/10.21098/jimf.v5i4.1098

Awad T, Awwad RA, Hamdan S (2025) The mediating role of governmental responsibilities and regulations toward corporate social responsibility: comparative study

Balarabe A, Abdullah MF, Gwadabe UM, Muhammad AJ (2023) History of Islamic banking in Malaysia: a general review. Int J Islamic Prod Malay Civ 2(1):43–53

DDCAG Group (2024) Exclusive Deepdive—Malaysia's Islamic Finance Market 2023

Faizi F (2023) Sharia governance framework in Islamic banking and finance: comparative analysis between Malaysia and Indonesia. Available at SSRN: https://ssrn.com/abstract=4622070 or https://doi.org/10.2139/ssrn.4622070

FitchRating (2025) Indonesian takaful sector transitions to higher capital rules. Available at: https://www.fitchratings.com/research/non-bank-financial-institutions/indonesian-takaful-sector-transitions-to-higher-capital-rules-09-07-2025

Ghanavati A et al (2024) Green growth and financial development in OECD economics. In: 2024 international conference on decision aid sciences and applications (DASA). IEEE

Haji-Othman Y, Yusuff MSS, Moawad AMK (2021) Analyzing zakat as a social finance instrument to help achieve the Sustainable Development Goals in Kedah. Econ Bus Stud 39(10):Article 5346. https://doi.org/10.25115/eea.v39i10.5346

Hamidi I, Fadillah A, Bashir A, Saputra A, Yarsah W (2025) Islamic financial inclusion and economic growth: empirical evidence from Indonesia. Indonesian Interdiscip J Sharia Econ (IIJSE) 8:2887–2902. https://doi.org/10.31538/iijse.v8i2.6069

IFSA 2013

INCEIF (2023) In a level playing field, risk-sharing contracts of Islamic finance can be a game changer. Available at: https://inceif.edu.my/kmimpact/2023/10/24/in-a-level-playing-field-risk-sharing-contracts-of-islamic-finance-can-be-a-game-changer/

Insurance Asia (2025) Malaysia's takaful industry poised for growth amidst regulatory reforms. Available at: https://insuranceasia.com/insurance/news/malaysias-takaful-industry-poised-growth-amidst-regulatory-reforms#:~:text=The%20general%20takaful%20sector%20recorded,compared%20to%20the%20previous%20year

Kuran T (1995) Islamic economics and the Islamic subeconomy. J Econ Perspect 9(4):155–173

OJK (2023) Indonesian Islamic Finance Report

Republic of Indonesia (2008) Law No. 21 of 2008 on Sharia (Islamic) Banking. Bank Indonesia

Rethink (2023) Malaysia, a Tiger Cub economy. Available at: https://rethinkq.adp.com/money-matters-malaysia/#:~:text=Malaysia's%20Bank%20Islam%20Chief%20Economist,banking%20system%2C%20Mohd%20Afzanizam%20says.

Saida D, Frikha M (2016) Banks and economic growth in developing countries: what about Islamic banks? Cogent Econ Finance 4:1168728. https://doi.org/10.1080/23322039.2016.1168728

Selangor (State) (2003) Administration of the Religion of Islam (State of Selangor) Enactment 2003 (Sel. Enactment No. 1 of 2003). Selangor State Legislative Assembly

Standard Chartered (2025) Standard Chartered: "Global Islamic Finance Assets to Surpass USD 7.5 Trillion by 2028. Available at: https://www.sc.com/en/press-release/standard-chartered-global-islamic-finance-assets-to-surpass-usd-7-5-trillion-by-2028/
World Bank (2020) Malaysia Islamic finance and financial inclusion

Driving Sustainable Innovation in Higher Education: Investigating the Behavioral Pathways to Artificial Intelligence Adoption in the United Arab Emirates

13

Hasan Yousef Aljuhmani, Razan Awadallah Awwad, Sameer Hamdan, and Hazim Aldabbas

Abstract

As higher education institutions worldwide strive to align with the United Nations Sustainable Development Goals (SDGs), the integration of emerging technologies such as artificial intelligence (AI) has become a key enabler of sustainable innovation. This study explores the behavioral pathways influencing AI adoption in public universities across the United Arab Emirates (UAE), a region at the forefront of digital transformation and educational sustainability. Anchored in a behavioral intention framework, the research investigates how performance expectancy, effort expectancy, facilitating conditions, attitude, and perceived risk shape academic staff's intention to adopt AI, with behavioral intention functioning as a mediating variable. Data were collected from 381 academic staff members using a simple random sampling method and analyzed using structural equation modeling. The results demonstrate that performance expectancy, effort expectancy, facilitating conditions, and attitude significantly impact behavioral intention, which in turn drives AI adoption. In contrast, perceived risk did not yield a significant effect. These findings highlight the need

H. Y. Aljuhmani
Department of Business Administration, Institute of Graduate Research and Studies, University of Mediterranean Karpasia, Lefkosa, Cyprus (Northern), Turkey

R. A. Awwad (✉)
Business and Economics, Al-Quds University, Jerusalem, Palestine
e-mail: Razan.awadallah@gmail.com

S. Hamdan
Business Administration, Near East University, TRNC, Turkey
e-mail: Sameer.hamdan@neu.edu.tr

H. Aldabbas
Faculty of Business and Economics, Girne American University, Kyrenia, Turkey

B. Awwad and R. A. Awwad (eds.), *Technology and Innovation Towards a Future of Business Sustainability*, Contributions to Environmental Sciences & Innovative Business Technology, https://doi.org/10.1007/978-981-95-6469-9_13

to address both technological enablers and user perceptions to promote sustainable AI integration in higher education. The study contributes to the ongoing discourse on digital innovation for sustainability, offering practical insights for institutional leaders and policymakers aiming to build agile, future-ready academic ecosystems.

Keywords

Artificial intelligence adoption • Behavioral intention • Higher education • Sustainable innovation • Digital transformation • UAE • Performance expectancy • Effort expectancy • Facilitating conditions • Perceived risk

13.1 Introduction

Artificial intelligence (AI) is rapidly emerging as a transformative force in the education sector, reshaping how students learn, instructors teach, and institutions operate. It has become one of the most sought-after technologies in today's academic landscape due to its potential to address persistent challenges such as high dropout rates, rigid assessment systems, textbook-dependent learning, and limited personalization in instructional design (Majid and Lakshmi 2020). The integration of AI is expected to significantly improve the quality, accessibility, and sustainability of education worldwide.

AI refers to the ability of machines to perform cognitive tasks traditionally associated with human intelligence, particularly learning and problem-solving (Chen et al. 2022). In recent years, advances in AI technologies have allowed for more dynamic and personalized educational experiences, enabling educators to tailor content and instruction to individual learner needs. As noted by Zhu and Ren (2022), AI applications in education are particularly evident through the use of information and communication technologies (ICT) in formal learning environments, allowing teachers to enhance engagement and instructional efficiency.

In the context of the UAE, AI adoption in higher education is gaining momentum as part of the country's broader digital transformation agenda. The use of AI holds the potential to improve decision-making, personalize student learning, and streamline academic governance within universities (Chatterjee and Bhattacharjee 2020; Awwad and Dhair 2025). Moreover, AI-powered knowledge systems can assist institutions in creating inclusive and sustainable educational ecosystems that support long-term development goals. While AI offers several promising tools for personalizing education (Kumar 2020; Awwad et al. 2023), its effective implementation still faces technical, organizational, and behavioral barriers.

Understanding the behavioral factors that influence AI adoption among academic staff is essential for fostering sustainable innovation in higher education. This chapter investigates the roles of performance expectancy, effort expectancy, facilitating conditions, attitude, and perceived risk in shaping behavioral intention toward AI adoption. Using the UAE as a case study, this research contributes to the growing body of knowledge on digital innovation in education by identifying the psychological and contextual enablers of sustainable technology integration. In doing so, it offers practical implications for university leaders and policymakers aiming to align digital strategies with educational sustainability goals.

13.2 Literature Review

AI is increasingly reshaping the landscape of education by enhancing instructional delivery, supporting administrative functions, and personalizing learning experiences. Its growing influence is evident across higher education institutions worldwide, where AI technologies are being leveraged to align with the goals of digital transformation and sustainability. In particular, the UAE has emerged as a proactive adopter of AI in education, reflecting national efforts to build a knowledge-based economy. AI adoption in higher education settings is not only a matter of technological readiness but also depends on a range of behavioral factors that influence individual acceptance and use of AI-enabled tools. This literature review synthesizes key constructs relevant to AI adoption: perceived risk, performance expectancy, effort expectancy, facilitating conditions, attitude, behavioral intention, and actual AI use in higher education.

13.2.1 Perceived Risk (PR)

Perceived risk remains a major barrier to the adoption of emerging technologies, particularly in digital service environments. Scholars offer varying definitions of perceived risk, but a common theme is the potential for negative outcomes. Featherman and Pavlou (2003) describe it as the possibility of loss or unfavorable consequences when using electronic services. In the educational context, Xie et al. (2021) extend this to the belief that using digital platforms could lead to unreliable or unsatisfactory results. Such concerns may hinder faculty members from fully embracing AI applications, especially when the outcomes of AI-based decision-making or content delivery are perceived as uncertain or misaligned with pedagogical goals.

13.2.2 Performance Expectancy (PE)

Performance expectancy (PE) is defined as the degree to which individuals believe that a given technology will improve their job performance (Williams et al. 2015). In the academic environment, this refers to educators' belief that AI can enhance teaching effectiveness, administrative efficiency, or student learning outcomes. As noted by Hutabarat et al. (2021), Awwad (2024), users are more inclined to adopt technologies that are perceived to add tangible value to their professional responsibilities. Abbad (2021) affirms that PE acts as a critical determinant of technology acceptance, while Miraz et al. (2022) highlight its significant predictive power in influencing adoption behaviors in educational settings.

13.2.3 Effort Expectancy (EE)

Effort expectancy (EE) concerns the perceived ease of use of a technology. It reflects the extent to which individuals believe that using a particular system will be free of effort and time-consuming processes. Technologies perceived as user-friendly are more likely to be adopted, especially in environments where educators manage multiple responsibilities. According to Samsudeen and Mohamed (2019), the ease of integrating AI tools into existing workflows positively influences users' intentions. In academic institutions, where resistance to change is often linked to technological complexity, EE plays a central role in adoption decisions.

13.2.4 Facilitating Conditions (FC)

Facilitating conditions (FC) refer to the extent to which individuals believe that technical and organizational infrastructure exists to support technology use (Ambarwati et al. 2020; Awwad and Hamdan 2015). In higher education, this includes the availability of IT support, training programs, funding, and clear policies for technology integration. Kamaghe et al. (2020) argue that inadequate support systems, delayed assistance, or lack of resources can significantly impede technology uptake. Therefore, institutional readiness and investment in support mechanisms are essential to ensuring successful AI integration (Awwad 2025).

13.2.5 Attitude (ATT)

Attitude toward technology reflects an individual's overall evaluation of using a system or engaging in a specific behavior. It is shaped by personal beliefs, prior experiences, and social influences. Gázquez-Abad et al. (2011) emphasize that attitudes are influenced by factors such as peer recommendations, institutional culture, and perceived usefulness. A positive attitude enhances users' willingness to experiment with new tools and persist in learning how to use them effectively. As Khurana et al. (2020) note, attitude directly influences technology acceptance by shaping how individuals assess and decide on the adoption of digital innovations.

13.2.6 Behavioral Intention (BI)

Behavioral intention (BI) represents an individual's readiness to perform a specific action, such as adopting a new technology. It serves as a strong predictor of actual behavior and is influenced by constructs such as PE, EE, ATT, and PR. According to Wang et al. (2023), BI is the internal motivation that drives individuals toward action. In the context of higher education, faculty members who exhibit strong behavioral intentions are more likely to incorporate AI tools into their teaching and administrative tasks, thereby supporting digital and sustainable innovation (Hamdan et al. 2021; Razzaque et al. 2019).

13.2.7 Adoption of AI in Higher Education (AAHE)

The adoption of AI in higher education has accelerated with the advancement of computing technologies and data-driven decision-making. As Ouyang and Jiao (2021) observe, AI is now widely used in AI in education applications, including intelligent tutoring systems, adaptive learning platforms, and educational analytics dashboards. These tools provide opportunities for enhancing personalized learning, increasing student engagement, and improving instructional quality (Razia and Awwad 2022; Bahaa and Bahaa 2024). According to Chen et al. (2020), such innovations offer capabilities beyond traditional teaching methods. Holmes et al. (2018) and Hwang et al. (2020) further argue that AI supports scalable and sustainable pedagogical models, making it an essential component of future-ready academic institutions.

13.3 Hypothesis Development

Understanding the behavioral dynamics that influence the adoption of AI in higher education is critical for ensuring the sustainability of digital transformation initiatives. Informed by behavioral theories and technology acceptance models, this section outlines the conceptual relationships between key determinants—such as perceived risk, performance expectancy, effort expectancy, facilitating conditions, and attitude—and behavioral intention, as well as their link to actual AI adoption. The following hypotheses are proposed based on a synthesis of relevant literature.

13.3.1 Perceived Risk (PR) and Behavioral Intention (BI)

Perceived risk (PR) is frequently identified as a critical inhibitor of technology adoption. In the context of online engagement, individuals often associate digital services with a higher level of uncertainty due to the absence of face-to-face interaction or tangible experiences. For instance, in e-commerce, perceived risk relates to concerns over product authenticity and financial security. Similarly, in education, faculty members may hesitate to adopt AI tools if they believe these technologies may lead to unreliable outcomes or compromise data privacy (Featherman and Pavlou 2003; Xie et al. 2021). Although studies have documented the negative influence of PR on online consumer behavior, findings remain mixed and context-dependent. Ha (2020) noted inconsistent correlations between PR and behavioral intention in digital environments. Therefore, in this study, PR is examined as a potential barrier to AI adoption in the UAE's higher education context.

Hypothesis 1 (H1): Perceived risk (PR) negatively influences behavioral intention (BI) to adopt AI in higher education.

13.3.2 Performance Expectancy (PE) and Behavioral Intention (BI)

Performance expectancy (PE) refers to the extent to which individuals believe that the use of a technology will enhance their job performance. In academic settings, this may involve improving teaching efficiency, streamlining administrative tasks, or enriching student learning outcomes (Williams et al. 2015). Prior research confirms that PE is one of the strongest predictors of technology adoption (Abbad 2021). For instance, Yu et al. (2021) demonstrated that PE, along with effort expectancy and enabling conditions, significantly influenced behavioral intention in the context of mobile health education. Alghamdi et al. (2022) also found that PE played a decisive role in students' acceptance of e-learning systems. These findings imply that when academic staff perceive AI technologies as beneficial for their educational duties, they are more likely to develop strong intentions to adopt them (Razia et al. 2022).

Hypothesis 2 (H2): Performance expectancy (PE) positively influences behavioral intention (BI) to adopt AI in higher education.

13.3.3 Effort Expectancy (EE) and Behavioral Intention (BI)

Effort expectancy (EE) is defined as the perceived ease of using a particular technology. In higher education, where faculty members are often burdened with multiple responsibilities, the usability of AI tools is critical. If the technology is perceived as complicated or time-consuming, educators may resist its integration into their daily routines. Studies such as those by Samsudeen and Mohamed (2019) and Yu et al. (2021) have shown that EE influences behavioral intention, although the strength of this relationship may vary by context. In some cases, EE had limited predictive power compared to other variables, suggesting that familiarity with digital tools or institutional support may moderate its effects. In the present study, EE is hypothesized to positively affect the behavioral intention to adopt AI in higher education.

Hypothesis 3 (H3): Effort expectancy (EE) positively influences behavioral intention (BI) to adopt AI in higher education.

13.3.4 Facilitating Conditions (FC) and Behavioral Intention (BI)

Facilitating conditions (FC) represent the perceived availability of organizational and technical infrastructure necessary to support technology use. This includes access to digital resources, IT support, training opportunities, and institutional encouragement. Abdullah et al. (2020) reported a significant relationship between FC and behavioral intention, especially when combined with factors such as social influence and hedonic motivation. Similarly, Azizi et al. (2020) highlighted that FC, along with social support and outcome expectancy, plays a vital role in influencing users' intent to engage with digital learning environments. In this study, FC is expected to positively influence behavioral intention, especially within institutions that actively invest in infrastructure and professional development for AI integration.

Hypothesis 4 (H4): Facilitating conditions (FC) positively influence behavioral intention (BI) to adopt AI in higher education.

13.3.5 Attitude (ATT) and Behavioral Intention (BI)

Attitude (ATT) refers to an individual's overall evaluation of performing a particular behavior, such as using a new technology. Attitudes are shaped by past experiences, perceived usefulness, ease of use, and social influence. A favorable attitude typically increases the likelihood of behavioral intention and eventual adoption. According to Ajzen's (1991) Theory of Planned Behavior (TPB), attitude is one of the key determinants of intention, along with subjective norms and perceived behavioral control. Wang et al. (2023) reinforce that attitudes reflect broader psychological inclinations that govern behavior. Mohamed Riyath et al. (2022) further confirm this relationship in the educational context, showing that students' positive attitudes toward platforms like Zoom significantly influenced their intention to use them. Thus, it is hypothesized that academic staff with positive attitudes toward AI will exhibit stronger intentions to adopt such technologies.

Hypothesis 5 (H5): Attitude (ATT) positively influences behavioral intention (BI) to adopt AI in higher education.

13.3.6 Behavioral Intention (BI) and AI Adoption in Higher Education (AAHE)

Behavioral intention (BI) is widely regarded as the most direct predictor of actual technology use. According to Chatterjee et al. (2021), individuals who express a clear intention to use AI are more likely to incorporate it into their practices. Almaiah et al. (2021) emphasized that a supportive technological environment enhances the transition from intention to action, particularly in mobile learning systems. Adequate access to infrastructure—such as reliable internet, computing devices, and user-friendly interfaces—is vital to bridging this gap. Furthermore, Al-Adwan et al. (2023) found that students with high personal innovativeness and self-efficacy were more inclined to accept AI-based learning systems, viewing technological challenges as opportunities for growth. These insights suggest that behavioral intention plays a critical mediating role in translating psychological readiness into actual AI adoption (Fig. 13.1).

Hypothesis 6 (H6): Behavioral intention (BI) positively influences the adoption of AI in higher education.

13.4 Methodology

This study adopts a quantitative research design to investigate the behavioral pathways influencing AI adoption among academic staff in public universities in the UAE. The target population consists of 3245 academic staff members employed at public institutions, as reported by the UAE Ministry of Education (Ahmad et al.

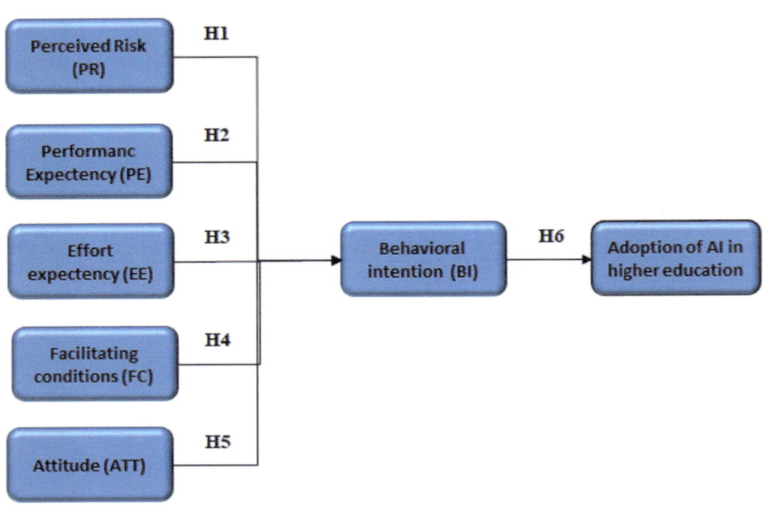

Fig. 13.1 Research model

2021). To determine an appropriate sample size, the RaoSoft calculator was used, suggesting a minimum of 344 participants at a 95% confidence level with a 5% margin of error.

To ensure sufficient representation, 400 questionnaires were distributed using a simple random sampling method. A total of 387 completed questionnaires were returned, yielding a high response rate of 96%. After screening for incomplete responses, 381 valid cases were retained for analysis. SPSS Version 27 was used for analyzing demographic data, while SmartPLS 4 was utilized to assess the measurement and structural models.

13.4.1 Data Analysis and Empirical Findings

Construct Reliability

The reliability of the measurement model was assessed using four standard indicators: Cronbach's alpha, composite reliability (CR) via rho_A and rho_C, and Average Variance Extracted (AVE). According to the guidelines by Hair et al. (2019), values above 0.70 for alpha and CR, and above 0.50 for AVE, indicate acceptable reliability and convergent validity. As reported in Table 13.1, all constructs met these criteria, confirming the reliability and internal consistency of the measurement scales.

Discriminant Validity

Discriminant validity was evaluated using the Heterotrait-Monotrait Ratio (HTMT), which assesses the distinctiveness of latent variables (Hair et al. 2019). Based on the threshold of 0.90 recommended by Henseler et al. (2015), all HTMT values were within acceptable limits (as shown in Table 13.1), confirming that the constructs exhibit sufficient discriminant validity.

Table 13.1 Reliability coefficients

Factors	Cronbach's alpha	CR (rho_a)	CR (rho_c)	(AVE)
AAHE	0.878	0.865	0.919	0.735
ATT	0.775	0.777	0.871	0.617
BI	0.863	0.873	0.911	0.684
EE	0.829	0.865	0.875	0.617
FC	0.830	0.864	0.877	0.623
PE	0.869	0.873	0.914	0.713
PR	0.926	0.926	0.931	0.789

13.4.2 Path Coefficient Testing

The structural model was evaluated using the bootstrapping procedure in Smart-PLS 4 with 5000 resamples to determine the significance of the hypothesized relationships. The results, summarized in Table 13.2, indicate that four of the six hypotheses were statistically supported.

Specifically, perceived risk (PR) had a significant positive effect on behavioral intention ($\beta = 0.177$, $p = 0.011$), indicating that concerns about risk do not necessarily deter adoption but may reflect cautious engagement. Performance expectancy (PE) also positively influenced behavioral intention ($\beta = 0.138$, $p = 0.037$), supporting the idea that perceived usefulness is key to technology acceptance. Attitude (ATT) emerged as a strong predictor of behavioral intention ($\beta = 0.212$, $p < 0.001$), while behavioral intention (BI) significantly predicted AI adoption in higher education ($\beta = 0.737$, $p < 0.001$), highlighting its central mediating role.

However, effort expectancy (EE) did not significantly influence behavioral intention ($\beta = 0.085$, $p = 0.276$). This suggests that perceptions of ease of use were not a critical determinant in this context, possibly due to the high digital literacy among UAE academic staff. Similarly, facilitating conditions (FC) showed no significant impact on behavioral intention ($\beta = 0.065$, $p = 0.439$), implying that access to resources and institutional support may already be sufficiently available and thus not a differentiating factor in influencing intent.

These findings validate the behavioral intention framework applied in the study and reveal nuanced insights into the enablers and barriers of AI adoption in higher education. Full empirical results are presented in Table 13.1 and Fig. 13.2. The next section discusses the theoretical and practical implications of these results in the broader context of sustainable innovation and educational transformation in the UAE.

Table 13.2 Test of path coefficients

Path	Relationships	β	S.E.	t-values	p-values	Decision
H1	PR → BI	0.177	0.069	3.173	0.011	Supported
H2	PE → BI	0.138	0.074	1.978	0.037	Supported
H3	EE → BI	0.085	0.080	1.078	0.276	Not supported
H4	FC → BI	0.065	0.082	0.767	0.439	Not supported
H5	ATT → BI	0.212	0.067	3.537	0.000	Supported
H6	BI → AI	0.737	0.017	28.024	0.000	Supported

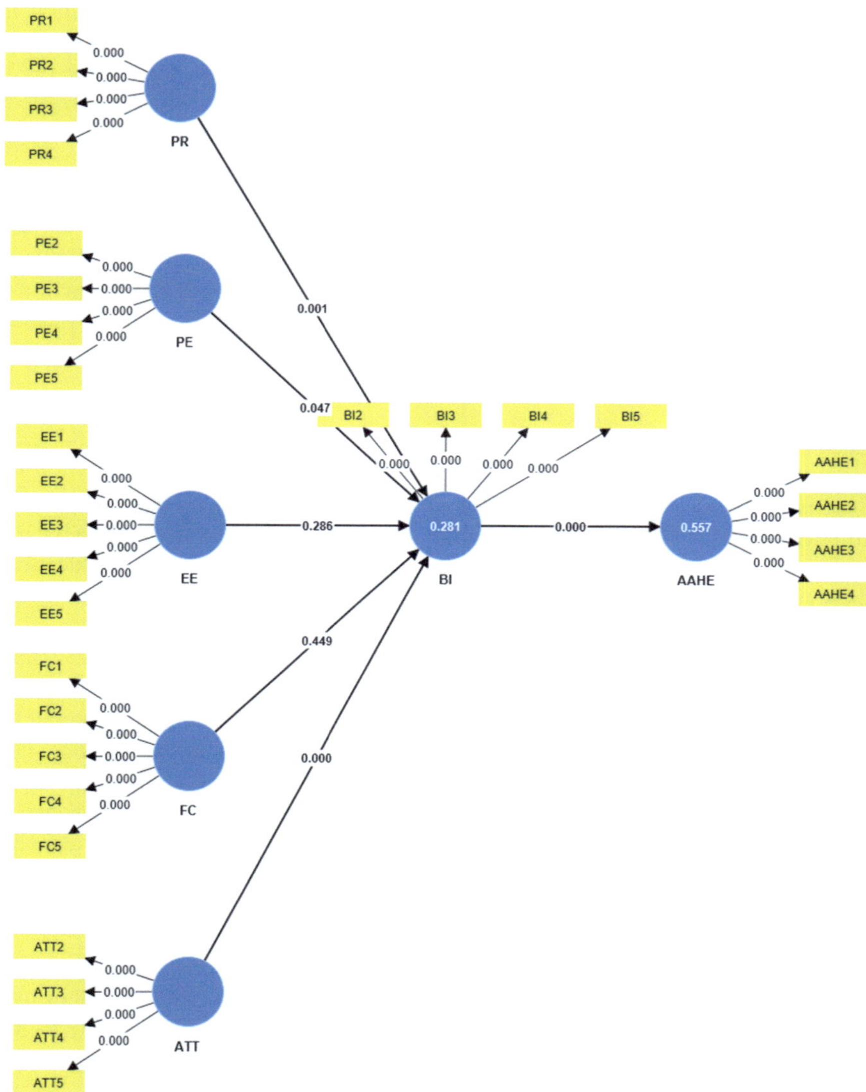

Fig. 13.2 Structural model results

13.5 Conclusion and Implications

The integration of AI in higher education is rapidly reshaping the academic landscape, with the potential to enhance teaching quality, administrative efficiency, and student engagement. As technological advancements continue to evolve, the need to understand the behavioral factors influencing AI adoption becomes increasingly urgent—especially in regions such as the UAE, where digital transformation is central to national development strategies. This study contributes to this understanding by examining how perceived risk (PR), performance expectancy (PE), effort expectancy (EE), facilitating conditions (FC), and attitude (ATT) affect behavioral intention (BI), which in turn drives the actual adoption of AI in higher education (AAHE).

The findings offer both theoretical and practical insights into the behavioral mechanisms shaping AI acceptance among academic staff at UAE public universities. Drawing from data collected through 381 validated responses and analyzed using structural equation modeling (SEM), the study tested six hypotheses and confirmed four, offering robust empirical evidence that enhances existing models of technology adoption in education.

13.5.1 Summary of Key Findings

The results indicate that perceived risk (H1), performance expectancy (H2), and attitude (H5) have significant positive effects on behavioral intention, while behavioral intention (H6) strongly predicts the actual adoption of AI in higher education. These findings are consistent with the broader literature, which emphasizes the role of trust, utility, and user mindset in shaping technology acceptance.

Specifically, perceived risk—traditionally assumed to negatively impact intention—was found to have a positive effect in this context, suggesting that academic staff are not deterred by risk but may engage with AI cautiously and deliberately. Performance expectancy, which reflects the perceived usefulness of a system, also emerged as a significant determinant of intention, reaffirming that users are motivated when they believe that technology can enhance their effectiveness. Likewise, attitude, defined as an individual's overall evaluative disposition toward technology, had the strongest influence on behavioral intention. This aligns with previous research suggesting that positive attitudes foster greater openness to adopting innovative systems (Chai et al. 2021).

In contrast, the study did not find support for effort expectancy (H3) or facilitating conditions (H4). Effort expectancy, which assesses the ease of system use, had no significant effect on behavioral intention. This may be attributed to the high digital fluency of faculty members in UAE universities, where training and exposure to educational technologies are widespread (Tannady and Dewi 2024). Similarly, facilitating conditions—defined as the institutional support and infrastructure available for using technology—did not significantly shape intention. It is likely that in technologically mature environments such as the UAE, access to resources and institutional backing is perceived as a given, and thus does not directly influence individual motivation (Kamaghe et al. 2020).

13.5.2 Theoretical Implications

This study makes several theoretical contributions to the field of educational technology adoption. First, it extends the Unified Theory of Acceptance and Use of Technology (UTAUT) by reaffirming the mediating role of behavioral intention in linking psychological predictors to actual system usage. While previous research has often emphasized effort expectancy and facilitating conditions as key predictors, the present study highlights the primacy of performance expectancy, attitude, and even perceived risk in high-tech educational contexts (Williams et al. 2015). This shifts the emphasis from ease of use and support availability toward perceptions of usefulness, trust, and value alignment.

Moreover, the findings suggest that attitude, often overlooked in favor of more operational constructs, remains a critical predictor of intention in education-related settings. Incorporating attitude into AI adoption frameworks in future research could improve the explanatory power of such models. The study also contributes by examining AI use in a non-Western context, providing fresh empirical insights that add cultural and institutional diversity to existing knowledge.

13.5.3 Practical Implications

From a practical standpoint, the results offer several important lessons for policymakers, university administrators, and educational technology developers. First, initiatives aimed at enhancing faculty adoption of AI tools should prioritize demonstrating the performance benefits of such technologies. Training programs should emphasize how AI can improve pedagogical practices, streamline tasks, and enrich the student learning experience.

Second, institutions should invest in building positive attitudes toward AI through awareness campaigns, professional development workshops, and success stories from early adopters. Such interventions can foster a culture of innovation and reduce psychological resistance to technological change. While infrastructure and technical support remain essential, this study suggests that motivational and perceptual factors may play a more central role in adoption decisions—especially in digitally advanced institutions.

Lastly, the finding that perceived risk does not deter adoption implies that faculty may be open to experimentation as long as risks are acknowledged and managed transparently. Administrators should therefore frame AI policies with clear ethical guidelines and data protection measures to maintain trust and engagement.

13.5.4 Limitations and Directions for Future Research

Despite its contributions, this study has several limitations that future research should address. First, the use of a cross-sectional survey limits the ability to draw causal conclusions. Longitudinal studies are recommended to examine how attitudes and behavioral intentions toward AI evolve over time, especially as exposure and experience increase.

Second, the study relied solely on quantitative, self-reported data, which may be subject to social desirability bias. Future studies should incorporate qualitative methods such as interviews or focus groups to gain richer insights into faculty experiences with AI. Triangulating survey data with observational or usage data would also enhance the validity of findings.

Third, while the sample was adequate for analysis, its generalizability is limited to UAE public universities. Expanding the sample to include private institutions or comparative studies across countries would improve the applicability of the results. Additionally, future models could include new constructs—such as perceived ethical risks, trust in AI algorithms, or digital identity—that reflect emerging dimensions of AI acceptance (Ashlin Nimo and Ravishankar 2024).

In conclusion, this study provides a timely and evidence-based perspective on how AI adoption in higher education can be strategically guided by understanding behavioral intention. By focusing on what motivates or hinders academic staff in embracing AI technologies, institutions can more effectively align their digital transformation efforts with sustainable innovation goals. The findings offer a valuable framework for future research, policy formulation, and institutional capacity building in AI-enabled education.

References

Abbad MM (2021) Using the UTAUT model to understand students' usage of e-learning systems in developing countries. Educ Inf Technol 26(6):7205–7224

Abdullah N, Redzuan F, Daud NA (2020) E-wallet: factors influencing user acceptance towards cashless society in Malaysia among public universities. Indones J Electr Eng Comput Sci 20(1):67–74

Ahmad AR, Alhammadi AHY, Jameel AS (2021) National culture, leadership styles and job satisfaction: an empirical study in the United Arab Emirates. J Asian Financ Econ Bus 8(6):1111–1120

Ajzen I (1991) The theory of planned behavior. Organ Behav Hum Decis Process 50(2):179–211

Al-Adwan AS, Li N, Al-Adwan A, Abbasi GA, Albelbisi NA, Habibi A (2023) Extending the technology acceptance model (TAM) to predict university students' intentions to use metaverse-based learning platforms. Educ Inf Technol 28(11):15381–15413

Alghamdi AM, Alsuhaymi DS, Alghamdi FA, Farhan AM, Shehata SM, Sakoury MM (2022) University students' behavioral intention and gender differences toward the acceptance of shifting regular field training courses to e-training courses. Educ Inform Technol 27(1):451–468

Almaiah MA, Al-Lozi EM, Al-Khasawneh A, Shishakly R, Nachouki M (2021) Factors affecting students' acceptance of mobile learning application in higher education during COVID-19 using Ann-Sem modelling technique. Electronics 10(24):3121

Ambarwati R, Harja YD, Thamrin S (2020) The role of facilitating conditions and user habits: a case of Indonesian online learning platform. J Asian Financ Econ Bus 7(10):481–489

Ashlin Nimo JR, Ravishankar K (2024) Examining the adoption of artificial intelligence in higher education using UTAUT model

Awwad B (ed) (2024) The AI revolution: driving business innovation and research. Springer

Awwad BS (ed) (2025) Sustainability in light of governance and artificial intelligence applications. Emerald Group Publishing

Awwad BS, Dhair MA (2025) The impact of implementing digital transformation on the quality of government service provision in the Palestinian Ministry of Higher Education and Scientific Research. In: Sustainability in light of governance and artificial intelligence applications, p 71

Awwad B, Hamdan A (2015) Higher education governance and its impact on the quality of scientific research: evidence from Palestine and Bahrain. In: The fifth international Arab conference to ensure the quality of higher education

Awwad B, Anaewah M, Razia B, Salameh M (2023) Governance boards of trustees: quality of higher education and the outputs of scientific research. J Bus Socio-Econ Dev 4(1):49–65

Azizi SM, Roozbahani N, Khatony A (2020) Factors affecting the acceptance of blended learning in medical education: application of UTAUT2 model. BMC Med Educ 20:1–9

Bahaa SR, Bahaa SA (2024) Exploring the practices and strategies of active learning in higher education. Int J Acad Res Educ Rev 12(2):27–29

Chai CS, Lin PY, Jong MSY, Dai Y, Chiu TK, Qin J (2021) Perceptions of and behavioral intentions towards learning artificial intelligence in primary school students. Educ Technol Soc 24(3):89–101

Chatterjee S, Bhattacharjee KK (2020) Adoption of artificial intelligence in higher education: a quantitative analysis using structural equation modelling. Educ Inf Technol 25:3443–3463

Chatterjee S, Rana NP, Khorana S, Mikalef P, Sharma A (2021) Assessing organizational users' intentions and behavior to AI integrated CRM systems: a meta-UTAUT approach. Inf Syst Front 1–15

Chen X, Xie H, Hwang GJ (2020) A multi-perspective study on artificial intelligence in education: grants, conferences, journals, software tools, institutions, and researchers. Comput Educ Artif Intell 1:100005

Chen L, Jiang M, Jia F, Liu G (2022) Artificial intelligence adoption in business-to-business marketing: toward a conceptual framework. J Bus Ind Mark 37(5):1025–1044. https://doi.org/10.1108/JBIM-09-2020-0448

Featherman MS, Pavlou PA (2003) Predicting e-services adoption: a perceived risk facets perspective. Int J Hum Comput Stud 59:451–474

Gázquez-Abad JC, Canniére MHD, Martínez-López FJ (2011) Dynamics of customer response to promotional and relational direct mailings from an apparel retailer: the moderating role of relationship strength. J Retail 87(2):166–181

Ha N (2020) The impact of perceived risk on consumers' online shopping intention: an integration of TAM and TPB. Manag Sci Lett 10(9):2029–2036

Hair JF, Risher JJ, Sarstedt M, Ringle CM (2019) When to use and how to report the results of PLS-SEM. Eur Bus Rev 31(1):2–24

Hamdan A, Hassanien AE, Khamis R, Alareeni B, Razzaque A, Awwad B (2021) Correction to: applications of artificial intelligence in business, education and healthcare. In: Applications of artificial intelligence in business, education and healthcare. Springer International Publishing, Cham, pp C1–C1

Henseler J, Ringle CM, Sarstedt M (2015) A new criterion for assessing discriminant validity in variance-based structural equation modeling. J Acad Mark Sci 43:115–135

Holmes W, Anastopoulou S, Schaumburg H, Mavrikis M (2018) Technology-enhanced personalised learning: untangling the evidence

Hutabarat Z, Suryawan IN, Andrew R, Akwila FP (2021) Effect of performance expectancy and social influence on continuance intention in OVO. J Manaj 25(1):125–140

Hwang G-J, Xie H, Wah BW, Gašević D (2020) Vision, challenges, roles and research issues of Artificial Intelligence in education. Comput Educ: Artif Intell 1:100001. https://doi.org/10.1016/j.caeai.2020.100001

Kamaghe J, Luhanga E, Kisangiri M (2020) The challenges of adopting M-learning assistive technologies for visually impaired learners in higher learning institution in Tanzania. Int J Emerg Technol Learn 15(1):140–151. https://doi.org/10.3991/ijet.v15i01.11453

Kumar M (2020) Artificial intelligence in education: current insights and future perspectives. Intern J Educ Technol High Educ 17(1):1–18. https://doi.org/10.1186/s41239-020-00211-1
Khurana A, Kumar VR, Sidhpuria M (2020) A study on the adoption of electric vehicles in India: the mediating role of attitude. Vision 24(1):23–34
Majid I, Lakshmi YV (2020) Artificial intelligence in education, vol 18, no 2, p 106
Miraz MH, Hasan MT, Rekabder MS, Akhter R (2022) Trust, transaction transparency, volatility, facilitating condition, performance expectancy towards cryptocurrency adoption through intention to use. J Manag Inf Decis Sci 25:1–20
Mohamed Riyath MI, Muhammed Rijah UL, Rameez A (2022) Students' attitudes on the use of Zoom in higher educational institutes of Sri Lanka. Asian Assoc Open Univ J 17(1):37–52
Ouyang F, Jiao P (2021) Artificial intelligence in education: the three paradigms. Comput Educ Artif Intell 2:100020
Razia B, Awwad B (2022) A comprehensive review of blockchain technology and its related aspects in higher education. In: Technologies, artificial intelligence and the future of learning post-COVID-19: the crucial role of international accreditation, pp 553–571
Razia B, Awwad B, Taqi N (2022) The relationship between artificial intelligence (AI) and its aspects in higher education. Dev Learn Organ Int J 37(3):21–23
Razzaque A, Al Hashimi M, Musleh A, Ahlam H (2019) Learning readiness when sharing knowledge while e-learning. Opción Rev Cienc Hum Soc (19):2930
Samsudeen SN, Mohamed R (2019) University students' intention to use e-learning systems: a study of higher educational institutions in Sri Lanka. Interact Technol Smart Educ 16(3):219–238
Tannady H, Dewi CS (2024) Exploring role of technology performance expectancy, application effort expectancy, perceived risk and perceived cost on digital behavioral intention of GoFood users. J Inf Dan Teknol 80–85
Wang C, Ahmad SF, Ayassrah AYBA, Awwad EM, Irshad M, Ali YA et al (2023) An empirical evaluation of technology acceptance model for artificial intelligence in E-commerce. Heliyon 9(8)
Williams MD, Rana NP, Dwivedi YK (2015) The unified theory of acceptance and use of technology (UTAUT): a literature review. J Enterp Inf Manag. https://doi.org/10.1108/JEIM-09-2014-0088
Xie J, Ye L, Huang W, Ye M (2021) Understanding FinTech platform adoption: impacts of perceived value and perceived risk. J Theor Appl Electron Commer Res 16(5):1893–1911
Yu CW, Chao CM, Chang CF, Chen RJ, Chen PC, Liu YX (2021) Exploring behavioral intention to use a mobile health education website: an extension of the UTAUT 2 model. Sage Open 11(4):21582440211055721
Zhu J, Ren C (2022) Analysis of the effect of artificial intelligence on role cognition in the education system. Occup Ther Int 2022:1–12. https://doi.org/10.1155/2022/1781662

14 How Entrepreneurs' Leadership Skills Affect the Growth of Small and Medium-Sized Enterprises in Palestine

Nidal Darwish, Razan Awadallah Awwad, Sameer Hamdan, and Rasha Hisham Obeid

Abstract

The objective of this study is to offer a comprehensive and in-depth analysis of the influence that the leadership abilities of entrepreneurs have on the expansion of small and medium-sized enterprises. The preliminary data is acquired using a questionnaire completed by employees of Palestinian SMEs. The sample size is 172 respondents. The study's findings reveal a significant relationship between entrepreneurs' leadership qualities (including technical skills, self-skills, human skills, management skills, and intellectual skills) and the growth of SMEs.

Keywords

Entrepreneurial leadership skills (ELS) • Self skills • Management skills • Artistic skills • Intellectual Skills • Human skills

14.1 Introduction

The expansion of SMEs is significantly impacted by entrepreneurial leadership. This is a result of the prior leader's experience and knowledge, which positively affects the quantity of market prospects being recognized. In order to capitalize on market opportunities, the entrepreneur actively employs formal HR approaches,

N. Darwish · R. A. Awwad · R. H. Obeid
Business and Economics, Al-Quds University, Jerusalem, Palestine
e-mail: ndarwish@staff.alquds.edu

S. Hamdan (✉)
Business Administration, Near East University, TRNC, Turkey
e-mail: Sameer.hamdan@neu.edu.tr

B. Awwad and R. A. Awwad (eds.), *Technology and Innovation Towards a Future of Business Sustainability*, Contributions to Environmental Sciences & Innovative Business Technology, https://doi.org/10.1007/978-981-95-6469-9_14

such as setting challenging yet motivating goals for himself and his employees. They argue that challenging goals inspire employees to perform better than less demanding ones, provided they receive sufficient commitment, feedback, and information, as entrepreneurs are capable of identifying opportunities and taking risks (Adudarko 2017; Fernando 2025).

The actual expansion of SMEs is quantified in terms of assets, sales, and employees (Lekhanya 2016; Elekwachi and Akenbor 2025). Entrepreneurs' growth objectives are prejudiced by their control level, need for achievement, self-efficacy, tolerance for uncertainty and ambiguity, and registration procedures for a business. Entrepreneurs who have no drive to expand their businesses are less likely to succeed.

However, leadership could be described as a means of influencing others within an institution to achieve the organization's goals. Leaders must, however, not overlook other situational aspects to manage, such as changes in the growth, corporate environment and resource limits. To address these challenges, leaders must increase their current resources. Leadership necessitates a broad range of skills, which can be classified into five categories: self-talents, technical abilities, managerial abilities, human capacities, and intellectual capacities (Abd Razak 2019).

Entrepreneurs, play a crucial role in societal development by establishing and operating small and medium-sized enterprises that create income while also contributing to improved performance and firm growth. An entrepreneur is defined by their passion for working, their ambition for firm growth and development, and their incredible work energy and drive for success. An entrepreneur possesses a robust level of self-confidence and other key entrepreneurial attributes that contribute to the business's growth, evolution, and long-term success (Qanswah 2014).

Given the substantial contribution that small and medium-sized enterprises (SMEs) make to the Palestinian economy, SMEs were able to meet societal demands for goods and services. In Palestine, SMEs make a significant contribution to both employment creation and the country's GDP (Awwad 2024) emphasizes the importance of governance and entrepreneurship in supporting economic growth in Palestine, enhancing the role of small and medium enterprises as a key driver of economic development and job creation.

The primary objective of this study is to gain a deeper understanding of the abilities of entrepreneurial leaders who can influence the expansion of SMEs in Palestine and to examine the role that entrepreneurial attributes play in the expansion of SMEs in Palestine.

14.2 Theory and Research Model

14.2.1 Entrepreneur's Leadership Skills

Four requirements must be met in order to have the chance to become an entrepreneur. First, it is important to have circumstances or opportunities to repurpose resources in order to generate revenue. Second, entrepreneurship necessitates individual distinctions, including the ability to identify information about opportunities or gain privileged access to specific individuals. Third, risk-taking is essential. Fourth, Marsal et al. (2024) highlights the importance of mentorship and financial support in enhancing entrepreneurial motivation among family members, which is an important actor in developing leadership and entrepreneurship skills. Organizing people and resources is necessary for the entrepreneurial process (Shane 2003; Rahman et al. 2024).

Numerous leadership theories have been developed to show why and how some individuals become exceptional leaders, as interest in leadership psychology has increased over the past century (Cherry 2019). While later leadership theories examined other factors, such as situational situations and skill levels, early ideas focused on the traits that set leaders away from followers. Although numerous unique theories of leadership have been established, most of them may be categorized into one of eight main groupings, as shown in Fig. 14.1.

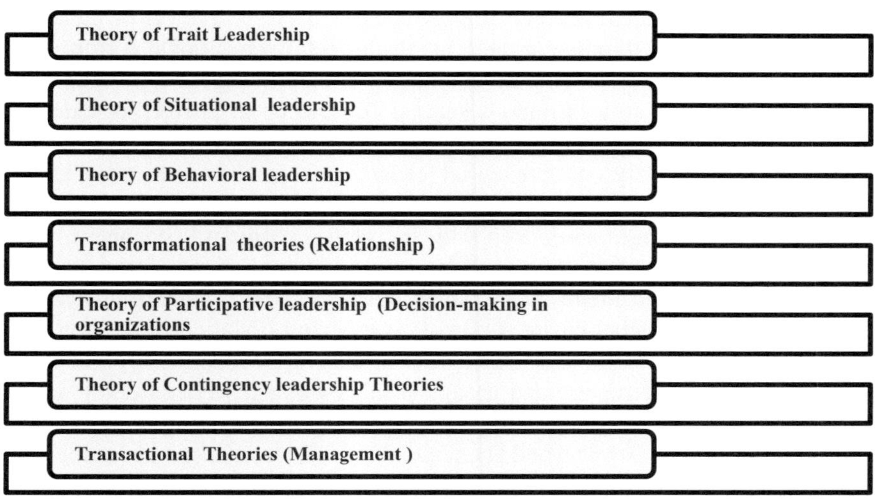

Fig. 14.1 Theories of entrepreneurial leadership

14.2.2 Growth of SMEs

Scholars have long emphasized the diversity of entrepreneurial ideas and said that no single theory can adequately explain how businesses thrive. This paradox has led to the development of numerous theories to explain different aspects of SMEs' expansion (Isaga 2012). Conferring to Abd Razak (2019), regularly evaluating the expansion and progress of projects from several perspectives, such as human resources and strategic locations, is essential to survival.

Growth demonstrates the investment and self-actualization of the entrepreneur in a personal business venture (Stam 2010). When the phenomena of gazelles, or high-growth firms, was discovered in the 1980s as those that could expand rapidly in a short amount of time, the importance of firm expansion to the macroeconomic system was first acknowledged (Birch and Medoff 1994; Acs et al. 2008).

14.2.3 The Relationship Between the Entrepreneur's Leadership Skills and SMEs' Growth: Hypothesis Development

More people are realizing how crucial strong leadership is to the expansion and prosperity of SMEs. Therefore, it is claimed that leadership and entrepreneurship are inextricably linked and that successful business people must be able to guide others (Cope and Kempster 2010). One of the most significant factors influencing organizational success is leadership, which has been the subject of numerous studies that have tried to pinpoint the traits that affect organizational performance. Therefore, business owners who are committed to employing the right leadership style may be the key to a business's success (Cascio et al. 2010). For example, Kieu (2010) found a strong correlation between profitability, sales growth, and leadership. Consequently, entrepreneurial leadership is viewed as a blend of human traits, abilities, and personalities (Sandybayev 2019).

Due to the challenges associated with the global economy's openness in the twenty-first century, the focus of the business world has switched to entrepreneurial leadership (Mishra and Misra 2017). Small business owners and executives must understand how the global business climate is evolving (Palalic 2017). Moreover, entrepreneurial leadership plays a vital role in enabling economic diversification and empowering underrepresented groups such as women, which are key drivers for sustainable economic growth, especially in regions transitioning from resource-dependent economies (Awwad and Hamdan 2023). This insight can be analogously applied to the Palestinian context, where fostering inclusive leadership can support SME expansion and broader socio-economic development. Entrepreneurial leaders, according to Greenberg et al. (2013), are those who are sufficiently familiar with themselves and their workplaces to grasp opportunities to create value for their companies, stakeholders, and society at large. Entrepreneurship theory posits that the personal traits of entrepreneurial leaders enable them to envision a prosperous future for their companies (Bagheri 2017). Through the identification and

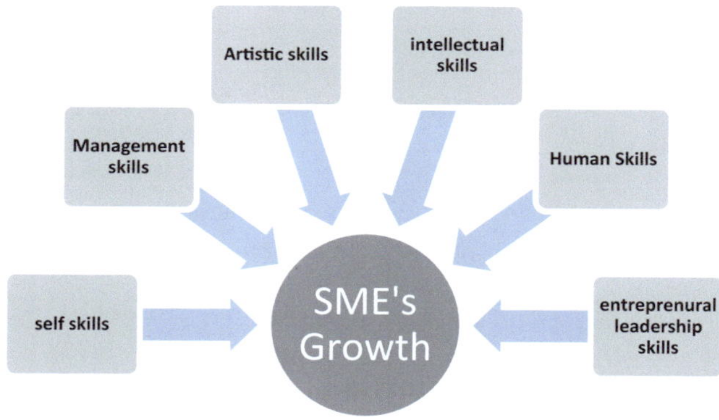

Fig. 14.2 Factors contributing to SME growth (adapted from Man et al. 2002)

exploitation of opportunities to enhance organizational performance, solve problems creatively, and effectively and efficiently use the organization's resources, entrepreneurial leadership supports and facilitates the adoption of procedures that strengthen the culture of organizational innovation (Rae 2017).

The talents and abilities reflected in the cognitive, mental, ethical, and psychological components that comprise the personality of officials in SMEs operating in the Bethlehem Governorate are referred to in this study as self-leadership skills. Physical prowess and physiological readiness are examples of self-skills, whereas leadership calls for concentration and a disciplined utilization of mental and physical resources (AL-Shboul 2025; Kanaan 2002) (Fig. 14.2).

14.3 Methodology

14.3.1 Overview of the Population (Palestine SMEs)

With more than 90% of enterprises and half of all jobs worldwide, SMEs play a critical role in both economic growth and employment. They make up 40% of the national GDP in emerging countries. The expansion of SMEs is a global priority, as it is estimated that 600 million jobs will be required by 2030 (World Bank 2022). SMEs play a vital role in local economic development by making substantial contributions to economic growth, poverty alleviation, and job creation (Gherghina et al. 2020).

According to the definition in Table 14.1 (MSMEs) encompass approximately 99% of all businesses operating in Palestine.

Table 14.1 Standards for classifying MSMEs in Palestine

Classification	Number of employees	Annual turnover
Micro	1–4	Turnover less than 100,000$
Very small	5–10	Turnover less than 200,000$
Small	11–20	Turnover less than 500,000$
Medium	21–50	Turnover less than 2000,000$
Large	50+	More than 2,000,000$

Source http://www.wafa.ps/Pages/Details/27197

14.3.2 Sample and Data Collection

The study population comprises SMEs that are active in Bethlehem's various economic sectors as of 2021. There are 8208 firms in the research population (Palestinian Central Bureau of Statistics 2020).

A total of 200 SMEs in the Bethlehem Governorate participated in the study, and 172 valid questionnaires were distributed. To improve communication, the survey was translated into Arabic. An 87.5% response rate was obtained from the 200 questionnaires that the researchers personally delivered, of which 175 were returned completely. Owners and managers of SMEs, as well as decision-makers in sales, marketing, and development, were the target respondents. SPSS was used to perform statistical analysis, which included correlations, means, frequencies, percentages, and standard deviations. The study aimed to address issues related to the expansion strategies of SMEs and test relevant theories.

14.4 Data Analysis

14.4.1 Descriptive Analysis

The researchers used the Pearson correlation coefficient to verify the internal consistency validity of the questionnaires and the sincerity of the arbitrators. To make sure the study tool produces accurate and trustworthy data, they also tested its stability using Cronbach's alpha coefficient. The stability coefficient of paragraphs, as determined by the split-half approach, was more than 0.7.

Table 14.2 indicates that all Cronbach's Alpha coefficients exceed 0.7, indicating that the instrument is stable and reliable.

14.4.2 Descriptive Statistics After Removing Outliers

Table 14.3 indicates that the stability coefficient of the nine SME growth items was 0.900. Additionally, this shows that the study instrument is appropriate for use in scientific research.

Table 14.2 Leadership skills dimensions Cronbach's alpha coefficients

	Dimension	Number of paragraphs	Cronbach's alpha coefficient
1	SLS	9	0.886
2	TLS	9	0.894
3	MLS	12	0.914
4	HLS	9	0.894
5	ILS	9	0.892
	Leadership skills	**48**	**0.971**

Table 14.3 Cronbach's alpha coefficients for the SMEs growth dimension

Dimension	Number of paragraphs	Cronbach's alpha coefficient
SMEs growth	9	0.900

Table 14.4 Correction key of the survey responses

Interval of the statistical mean	Description
1.00–1.79	Very low
1.80–2.59	Low
2.60–3.39	Moderate
3.40–4.19	High
4.20–5.00	Very high

Table 14.4 presents the responses extracted from the questionnaire paragraphs, which utilize a 5-point Likert scale (1–5). The range of responses is 4, indicating a high level. The range of responses is divided into 5 groups with a width of 0.8, which is considered the correction key for the mean of responses as shown below.

14.5 Limitations and Recommendations for Future Research

The researcher chose employees of SMEs in the Bethlehem Governorate to participate in a study on how entrepreneurial leadership abilities affect their development. After a considerable amount of time and effort, the questionnaires were completed, despite several attempts to do so.

To improve leadership abilities, develop effective business models, and instruct leadership skills courses, future research should focus on in-depth, sector-specific, and case study-based approaches.

14.6 Conclusions

According to the report, the development of SMEs in the Bethlehem Governorate is significantly influenced by leadership abilities. The greatest influence comes from intellectual leadership abilities, which are followed by management, technical leadership, self-leadership, and human leadership. Leaders with self-leadership skills can identify issues, anticipate changes, concentrate, make informed decisions, understand employee motivation, and effectively manage their emotions. On the other hand, executives with technical leadership abilities are responsible both publicly and privately, possess cultural awareness, set reasonable goals, and strike a balance between sound judgment and employee sentiment. In SMEs, these abilities boost output and worker productivity. The importance of these abilities in promoting the expansion of SMEs in the Bethlehem Governorate is emphasized by the study.

According to the findings, SMEs in the Bethlehem Governorate possess strong leadership and management capabilities. These abilities include efficiently managing time, gathering reliable information, and making informed decisions. These abilities help SMEs expand by increasing their market share and competitiveness. Effective communication, teamwork, collaboration, and mutual respect are all essential components of human leadership. By ensuring the best member selection and accurate task specification, these abilities also contribute to the formation of productive work teams. These abilities enable SMEs to establish institutional foundations for their businesses, ensuring their success for many generations to come.

Intellectual leadership abilities have a significant impact on the development of SMEs in the Bethlehem Governorate. These abilities empower leaders to understand issues, make informed decisions, and exhibit ethical behaviour. They also encourage creativity and staff loyalty. The effect of leadership abilities on the expansion of SMEs does not differ significantly by gender. However, the benefits of leadership abilities on the growth of SMEs vary at the 5% level of significance, depending on various parameters, including the number of employees, economic activity, company age, years of experience, job position, qualifications, and age.

References

Abd Razak D (2019) Learning to lead and strategic decision making in shared leadership organisations: challenges and prospects for start-ups SMEs' in Malaysia. Bournemouth University, Malaysia

Acs Z, Parsons W, Tracy S (2008) Highimpact firms: gazelles revisited, no 328. SPA Office of Advocacy, Washington

Adudarko G (2017) Entrepreneurial leadership and growth of small medium enterprises: lessons from transformational entrepreneurs in Ghana. University of Ghana, Ghana

AL-Shboul MA (2025) The correlations between business ethics rules, talented human resource supply chain management and managing SMEs ethics: fresh insight from middle Eastern countries. Int J Soc Econ 52(7):961–975

Awwad BSAL (2024) Governance with relationship between entrepreneurship and economic growth in Palestine. Int J Law Manag 66(2):259–287

Awwad BSAL, Hamdan AMM (2023) Saudi Arabia's post-oil economic prospects: economic diversification, entrepreneurship and women's empowerment. J Hum Resour Sustain Stud 11:903–941

Bagheri A (2017) The impact of entrepreneurial leadership on innovation work behavior and opportunity recognition in hightechnology SMEs. J High Technol Manag Res 28(2):159166

Birch D, Medoff J (1994) Gazelles. In: Solmon LC, Alec RL (eds) Labor markets, employment policy and job creation. Westview Press, Boulder, p 159168

Cascio WF, Boudreau JW (2010) Investing in people: financial impact of human resource initiatives (2nd ed). Pearson Education

Cherry K (2019) The major leadership theories: the 8 major theories of leadership. https://www.verywellmind.com/leadershiptheories2795323

Cope J, Kempster S (2010) Learning to lead in the entrepreneurial context. Hunter Centre for Entrepreneurship

Elekwachi AB, Akenbor LC (2025) Marketing intelligence system and sales of beverages by small and medium scale enterprises in Yenagoa local government area. BW Acad J 2

Fernando A (2025) Unlocking the path to international expansion: investigating the impact of managerial international experience on SME internationalization. FIRM J Manag Stud 8(2):274–306

Gherghina S, Botezatu M, Hosszu A, Simionescu L (2020) Small and medium sized enterprises (SMEs): the engine of economic growth through investments and innovation. Sustainability 12(347):122

Greenberg D, McKone-Sweet K, Wilson H (2013) Entrepreneurial leaders: creating opportunity in an unknowable world. Leader to Leader 2013(67):5662

https://www.worldbank.org/en/topic/smefinance

Isaga N (2012) Entrepreneurship and the growth of SMEs in the furniture industry in Tanzania. Vrije Universiteit, Tanzania

Kanaan N (2002) Administrative leadership. House of Culture Library, Amman, Jordan

Kieu H (2010) Leadership styles and organizational performance: a predictive analysis. University of Phoenix, USA

Lekhanya L (2016) Determinants of survival and growth of small and medium enterprises in rural KwaZulu—Natal. University of Western Cape, South Africa

Man TWY, Lau T, Chan KF (2002) The competitiveness of small and medium enterprises: a conceptualization with focus on entrepreneurial competencies. J Bus Ventur 17(2):123–142. https://doi.org/10.1016/S0883-9026(00)00058-6

Marsal H, Hamdan A, Awwad BSALA, Mohamed M (2024) The impact of mentorship and funding support on stimulating entrepreneurship motivation among family members. Eur J Fam Bus 14(1):117–130

Mishra P, Misra R (2017) Entrepreneurial leadership and organizational effectiveness: a comparative study of executives and nonexecutives. Procedia Comput Sci 122:71–78

Palalić R (2017) Entrepreneurial leadership and globalization: challenges and opportunities for SMEs. Springer

Palestinian Central Bureau of Statistics (2020) Statistical yearbook of Palestine 2020, no 21. Ramallah, Palestine

Qanswah A (2014) Role of enterprenureship in private installations performance efficiency in Concrete Engineering Company Ltd. In: Case study: Concrete Engineering Company Ltd. Sudan University of Science & Technology, Sudan

Rae D (2017) Developing entrepreneurial leadership: the challenge for sustainable organisations. Int J Work Innov 2(1):76–100

Rahman S et al (2024) Developing students' life skills through entrepreneurship and agropreneurship education at Al-Itqan Islamic Boarding School, Depok. J Learn Improv Lesson Study 4(1):9–24

Sandybayev A (2019) Impact of effective entreprenerial leadership style on organizational performance: critical review. Int J Econ Manag 1(1):47–55

Shane S (2003) A general theory of entrepreneurship: the individual opportunity nexus. E. Elgar, Northampton, Massachusetts

Stam E (2010) Growth beyond Gibrat: firm growth processes and strategies. Small Bus Econ 35(2):129135

World Bank (2022) Small and medium enterprises (SMEs) finance: improving SMEs' access to finance and finding innovative solutions to unlock sources of capital. https://www.worldbank.org

15 Unlocking Potentials: The Impact of English Language Proficiency on Employability and Job Opportunities in the Palestinian Market. VOSviewer AI Application

Razan Awadallah Awwad, Sameer Hamdan, and Fatima Y. Awwad

Abstract

This book chapter investigates the influence of English language proficiency on employability and job opportunities within the Palestinian labor market. English proficiency and English language teaching are increasingly recognized as a critical factor in enhancing graduates' competitiveness in accessing high-quality job vacancies. The study employed a comprehensive bibliometric review methodology, utilizing PRISMA guidelines and VOSviewer application to analyze relevant literature systematically. The originality of this research lies in its focus on the language-employment nexus in a developing country context, offering one of the first bibliometric analyses of the English language gap and its implications for recruitment and employability. The findings indicate that graduates with strong English language skills are more likely to secure better job opportunities and attain more prestigious positions.

Keywords

Employability • English language teaching • English proficiency • Job market requirement • Bibliometric analysis

R. A. Awwad
Business and Economics, Al-Quds University, Jerusalem, Palestine

S. Hamdan
Business Administration, Near East University, TRNC, Turkey
e-mail: Sameer.hamdan@neu.edu.tr

F. Y. Awwad (✉)
English Department, Birzeit University, Ramallah, Palestine
e-mail: fatima.q88@gmail.com

B. Awwad and R. A. Awwad (eds.), *Technology and Innovation Towards a Future of Business Sustainability*, Contributions to Environmental Sciences & Innovative Business Technology, https://doi.org/10.1007/978-981-95-6469-9_15

15.1 Introduction

The English language has become increasingly vital in a globalized world. In the Palestinian context, where unemployment among young people is high (Awwad et al. 2022), languages in general have become a vital asset for employment. This chapter investigates how ELT and English proficiency support student development, enhance employability, and open doors to wider opportunities in the marketplace (Nassar and Nassar 2025).

In Palestine, English has become a pivotal language in the labour market, particularly in sectors like business, technology, and international relations. As globalization integrates economies worldwide, English language is essential for employability and professional mobility. However, disparities exist in the quality of English language teaching between local and international institutions (Raymond 2025).

The slogan that English opens doors of opportunities for people who have different languages has echoed over the years as languages lead to several social, economic and educational opportunities (Adebola 2019).

Recent interest in employability research has included studies on the nexus between language proficiency and employability, as well as the value of English in employment opportunities, which is one of the most dominant topics (Huynh et al. 2023).

In an increasingly globalized labour market, English proficiency has become a crucial asset for job seekers worldwide, including Palestine. Recent graduates face high expectations from employers, who view English skills as a key factor for competitiveness, especially in sectors such as business, technology, and international NGOs (Nassar and Nassar 2025).

This paper highlights the importance of English skills for graduate job applications in various occupational sectors of non-native English-speaking countries (Nghia et al. 2023). Moreover, this chapter examines the impact of English proficiency on the employability and career prospects of Palestinian graduates, with a focus on local market opportunities and international exposure.

15.2 Literature Review

The link between English proficiency and employability has been widely documented. English is often considered a "passport" to better job opportunities worldwide (Melibari 2025; Awadallah et al. 2023). In Palestine, research indicates that graduates with strong English skills secure better-paying and more prestigious jobs, particularly in multinational companies, NGOs, and public sector roles (Palestinian Ministry of Education 2023; Adebola 2019; Melibari 2025).

Empirical research conducted in Palestine highlights the significant role of English language proficiency in enhancing employability and expanding career opportunities. For instance, studies at Hebron University demonstrate that the integration of computer-assisted language learning (CALL) improves English proficiency, motivation, and self-confidence (Alyan and Salameh 2022). Additionally, English proficiency among medical students is associated with better access to international resources and higher employability, despite some challenges such as academic pressure and communication barriers (Al-Tamimi and Al-Jabari 2022). Research in Malaysia has highlighted that a lack of English proficiency is one of the primary reasons contributing to the high unemployment rate among graduates in this country (Ting et al. 2017). And according to Al-Mahrooqi and Tuzlukova (2014), English has been viewed as an essential instrument for progress and the acquisition of knowledge and technology in Oman. Al-Tamimi and Shuib (2009), stated that the majority of petroleum engineering graduates who applied for jobs in Yemeni oil companies were rejected due to their weak English proficiency. Since businesses in the Middle East and North Africa favor candidates who speak English fluently, Erling (2015) asserted that proficiency in the language is acknowledged as significantly enhancing an individual's employability in these regions. According to the facts, knowing English is now essential for anyone hoping to advance in their job in a global setting.

Despite the recognized importance of English, high youth unemployment rates persist in Palestine, underscoring the need for continuous evaluation of educational policies aimed at improving employability through language skills (Salim 2022). Overall, these studies emphasize the multifaceted impact of English proficiency on employability, highlighting the importance of modern teaching methods, teacher training, and cultural sensitivity.

15.2.1 English Language and Employment Opportunity in Non-English-Speaking Countries

Proficiency in foreign languages has emerged as a critical component of people's employment opportunities. Although millions of people speak English today, many English speakers are not native to the language. English language competency is crucial for enhancing the employability of job seekers in non-native English-speaking countries (Al-Mahrooqi and Denman 2018; Doan and Hamid 2019).

15.2.2 English Language as a Job Market Requirement

Employers in Palestine, particularly, now consider English proficiency as a baseline skill. Job postings for positions are listed in English, and fluency in the language is considered essential or highly preferred in the hiring process. The increase in remote work and global outsourcing has further improved this demand (World Bank 2022).

15.2.3 Expectations from Employers

A study conducted by GIZ (2023) found that over **65%** of Palestinian employers prefer candidates with intermediate to advanced English skills. Key areas where English is required include writing reports and emails, participating in virtual meetings, and preparing presentations and proposals for projects.

Employers often test candidates' English ability during interviews, and many multinational companies in Palestine conduct interviews entirely in English (Massar Associates 2023).

Demand for English Skills is required the most in Information and Communication Technology (ICT), Education and Training, NGOs, International Agencies and Tourism (Adebola 2019).

15.2.4 Gaps Between Education and Market Needs

A common issue cited by employers is the mismatch between university English curricula and the demands of the workplace. Universities often focus on grammar and literature rather than practical, job-oriented communication skills (Zughoul 2023). Palestinian students studying abroad reportedly achieve higher English proficiency and better soft skills, improving their employability upon return (USAID 2023).

15.3 Methodology

Given the increasing importance of integrating employability with ELT and English proficiency in addressing labour market efficiency, a comprehensive bibliometric and systematic review approach was employed in the study, supported by data collection and analysis methodology using the Scopus database. To ensure that studies are directly related to the topic, a comprehensive search strategy was developed, focusing on titles, abstracts, and keywords (TITLE-ABS-KEY) using the terms "employability" and "English language teaching" or "English proficiency". A total of 143 initial results were found, and after filtering according to subject area, 127 remained. This targeted search ensures that the retrieved studies are highly relevant to the core focus of the investigation.

To ensure the quality of the data, strict filtering criteria were applied. Only publications in English were included, covering the period from 2015 to 2025 to encompass recent research. Moreover, the search was further restricted to subject areas only. After applying all filters and criteria, a total of 143 documents were retrieved from the database during the specified timeframe. The data were exported in CSV format, containing detailed information such as titles, authors, keywords, abstracts, and dates, which facilitated precise bibliometric analysis. The researcher conducted manual screening to ensure relevance and exclude off-topic or low-quality studies. As a result, the final dataset included only 52 studies that met the inclusion criteria and were suitable for in-depth qualitative synthesis under the PRISMA framework. See Fig. 15.1.

For data analysis, advanced bibliometric tools such as VOSviewer were employed to visualise scientific networks, including co-authorship networks, keyword co-occurrence, and citation patterns. Concurrently, a systematic review was conducted in accordance with PRISMA guidelines for qualitative content analysis. Studies were classified based on AI applications, research methodologies, and targeted environmental domains.

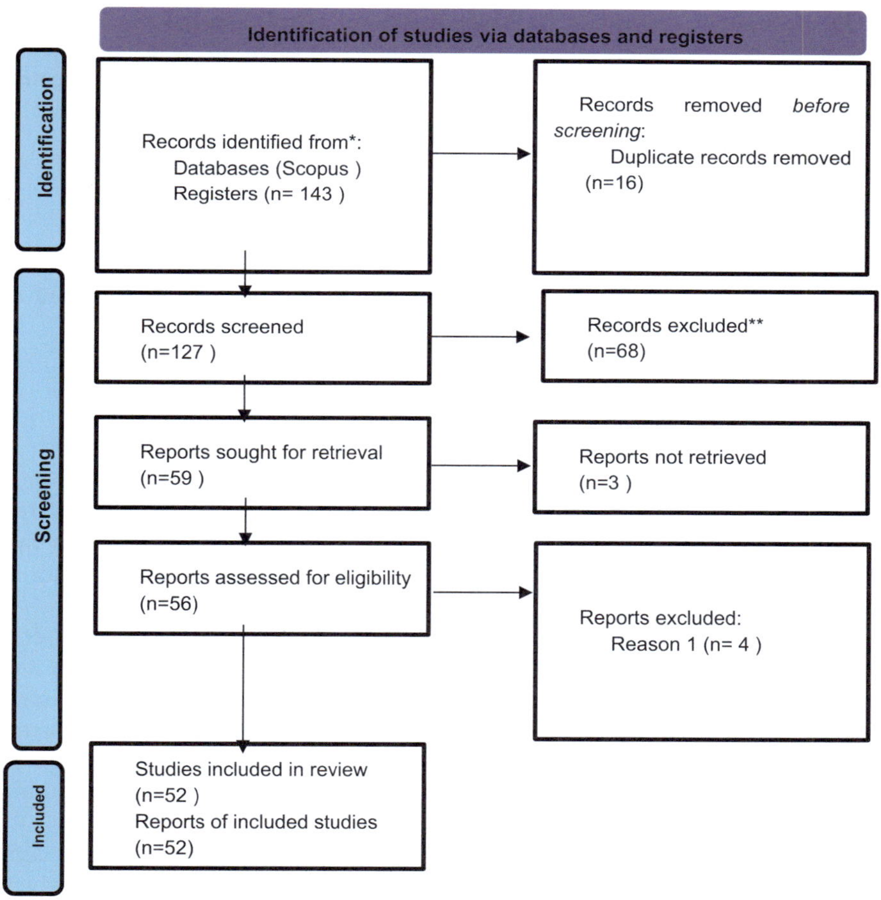

Fig. 15.1 The process of identification of studies

15.4 Discussion and Results

We conducted the study selection in accordance with PRISMA guidance, promoting transparency and methodological rigor at every step. Our search yielded 143 records from Scopus and complementary registry sources. After removing 16 duplicates, we obtained 127 distinct records for the initial screening. Title-and-abstract review led to the exclusion of 68 records for reasons of irrelevance or non-fulfillment of the predefined inclusion criteria. We then attempted to obtain the full texts of the remaining 59 reports; however, 3 could not be retrieved. Accordingly, 56 full-text articles underwent eligibility assessment. Of these, we excluded 4 studies because of inadequate methodological reporting or divergence from the scope of inquiry. In the end, 52 studies satisfied all inclusion requirements and were incorporated into the final analysis. This sequence of steps secured a concentrated and high-quality collection of studies for the review. See Fig. 15.1.

15.4.1 Classification of Studies by Research Co-occurrence Network

This network visualizes how titles, abstracts, or author keywords are related to each other based on co-occurrence in the selected research papers.

15.4.2 Classification of Studies by Research Keywords

The VOSviewer map illustrates the co-occurrence relationships among keywords in academic literature related to employability, specifically in the context of higher education and English language teaching. The size of each node reflects the frequency of keyword appearance, and the connecting lines indicate co-occurrence in the same documents. Different colours represent distinct thematic clusters.

The study reveals English proficiency among students educated abroad, correlating with superior employment outcomes (higher salary, job relevance) and Graduates from abroad report enhanced confidence, communication, and intercultural competencies, positively affecting their employability. Local graduates face challenges in securing jobs requiring advanced English skills due to gaps in ELT quality. Moreover, Employers emphasise English proficiency as a key hiring criterion, particularly in private and international sectors.

The co-occurrence network analysis generated five distinct clusters, each reflecting a unique aspect of the employability discourse. The green cluster, which occupies a central position in the map, emphasises the interrelationship between employability, graduate employability, communication skills, higher education, and English language proficiency. This cluster highlights the crucial role of linguistic and communicative competencies in higher education, enhancing graduate employability. The red cluster centers on employment, labour market, skills, education, and careers indicating a strong focus on the alignment between educational outcomes and labour market demands. The yellow cluster highlights the educational and instructional dimensions of employability, featuring keywords such as English language, students, English teaching and employers. This grouping underscores the importance of English language in preparing students to meet employer expectations. The blue cluster draws attention to the geopolitical and policy contexts of English language teaching, as evident in terms such as ELT, EFL, language policy, English as a medium of instruction, and country-specific references, including India, Vietnam, and Oman. This cluster points to the global nature of English language education and its implications for employability in non-native English-speaking countries. Finally, the purple cluster addresses more conceptual and analytical concerns, incorporating terms such as needs analysis, neoliberalism, Bangladesh, and placement. This cluster suggests a focus on theoretical frameworks and country-specific analyses, particularly those examining the influence of neoliberal educational reforms on employability strategies. Collectively, these clusters demonstrate the multidimensional and global character of research at the intersection of English language education and employability as shown in Fig. 15.2.

The visualized keyword reveals three interrelated clusters green, red, and blue highlighting the multifaceted nature of employability and other variables. The green cluster emphasises the strong connection between employability, English language proficiency, communication skills, and specific national contexts, such as Oman, suggesting that these are essential competencies for graduate success

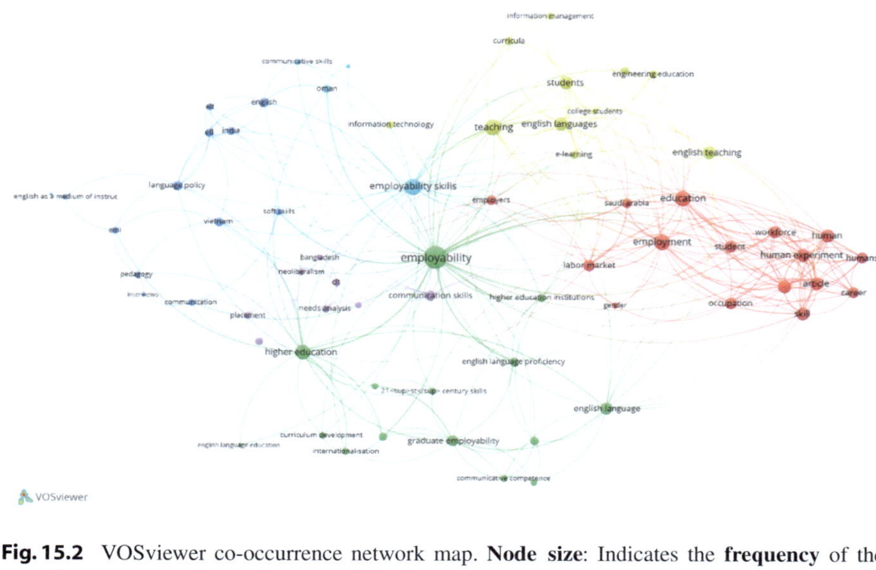

Fig. 15.2 VOSviewer co-occurrence network map. **Node size**: Indicates the **frequency** of the term (larger = more frequent). **Lines (Edges)**: Show **co-occurrence** relationships (terms appearing together). **Color clusters**: Represent **thematic groupings** of terms, identified by clustering algorithms

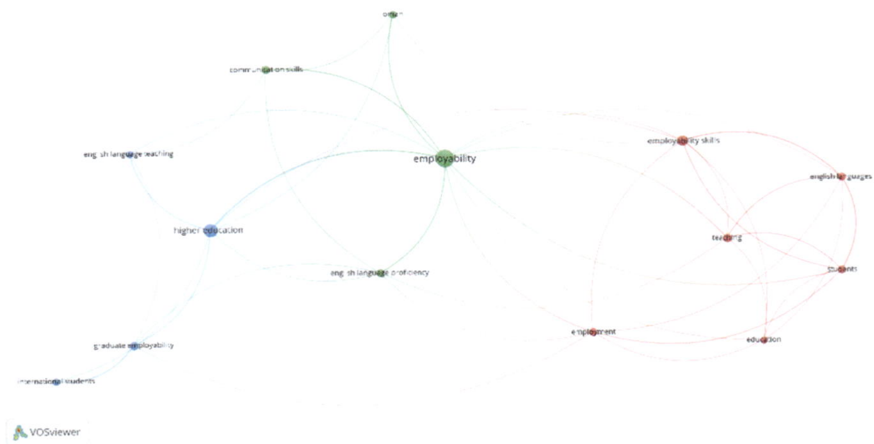

Fig. 15.3 VOSviewer co-occurrence network graph. **Node size**: Indicates the **frequency** of the term (larger = more frequent). **Lines (Edges)**: Show **co-occurrence** relationships (terms appearing together). **Color clusters**: Represent **thematic groupings** of terms, identified by clustering algorithms

and regionally influenced outcomes. The red cluster connects educational elements, such as teaching the English language, to the acquisition of employability skills, reflecting the role of formal education in labour market preparedness and the development of practical, job-related competencies. In contrast, the blue cluster emphasizes the significance of higher education institutions, particularly in international contexts, where English language teaching has a positive impact on employability. Overall, the graph highlights that employability is not isolated from language skills, personal competencies, and market expectations and opportunities, as reflected through the interplay of these three color-coded clusters, as shown in Fig. 15.3.

English language teaching and English proficiency in Palestine play a critical role in shaping students' futures. Beyond academic success, it enhances employability and creates new avenues for engagement in the local and global economy. Strengthening ELT through investment in teachers, curriculum reform, and technological support can significantly improve the prospects of Palestinian students and contribute to broader social and economic development.

English proficiency is no longer a luxury but a necessity for recent graduates in Palestine seeking employment in competitive sectors. Addressing the gap between academic preparation and market expectations is crucial for enhancing employability outcomes and promoting economic development.

In Palestine, research highlights that graduates with strong English skills secure better-paying and more prestigious jobs, particularly in multinational companies, NGOs, and public sector roles (Palestinian Ministry of Education 2023).

In summary, in our increasingly internationalized society, employment and English proficiency are related.

15.5 Recommendations for Future Research

Addressing the gap between English language education and employability in the Palestinian context requires a multi-level, collaborative approach. Future research can be conducted using a quantitative approach with a longitudinal design. Additionally, the study can be done using a governmental policy or the workplace environment as a mediator variable.

References

Adebola LO (2019) How well prepared are Chilean beginning English teachers to meet job demands in the English language teaching industry? J Entrep Educ 22(6):1–12

Almassri AN (2023) Learning for employability in contexts of economic despair: empirical insight from the Gaza Strip. J Teach Learn Graduate Employ 14(1):220–236. https://doi.org/10.21153/jtlge2023vol14no1art1711

Al-Mahrooqi R, Tuzlukova V (2014) English communication skills and employability in the Arabian Gulf: the case of Oman. Pertanika J Soc Sci Hum 22(2)

Al-Mahrooqi R, Denman C (eds) (2018) English education in Oman: current scenarios and future trajectories. English language education, vol 15. Springer Singapore. https://doi.org/10.1007/978-981-13-0265-7

Al-Tamimi A, Shuib M (2009) Motivation and attitudes towards learning English: a study of petroleum engineering undergraduates at Hadhramout University of Sciences and Technology. GEMA Online J Lang Stud 9(2):29–55

Al-Tamimi N, Al-Jabari M (2022) English language proficiency and its impact on employability among Palestinian medical students. PMC Central 14(3):Article 9348932. https://pmc.ncbi.nlm.nih.gov/articles/PMC9348932/

Alyan S, Salameh A (2022) The effectiveness of computer-assisted language learning (CALL) on English reading and writing skills of Palestinian university students. Hebron Univ J 5(2):78–92. https://digitalcommons.aaru.edu.jo/hujr_b/vol5/iss2/11/

Awadallah R, Hamdan S, Assi AF (2023) Work-family conflict's influence on job satisfaction. A mediator role of emotional exhaustion. In: 2023 international conference on sustainable Islamic business and finance (SIBF). IEEE, pp 89–94

Awwad RI, Aljuhmani HY, Hamdan S (2022) Examining the relationships between frontline bank employees' job demands and job satisfaction: a mediated moderation model. Sage Open 12(1):21582440221079880

Deutsche Gesellschaft für Internationale Zusammenarbeit (GIZ) GmbH (2023) Programme for Access to the Labour Market (PALM II) [Report]. GIZ. https://www.giz.de/en/downloads/giz-2023-en-PALM.pdf

Doan LD, Hamid MO (2019) Economics of English: examining the demand for English proficiency in the Vietnamese job market. RELC J 52(3):618–634. https://doi.org/10.1177/0033688219863166

Erling E (2015) The relationship between English and employability in the Middle East and North Africa. http://oro.open.ac.uk/44825/

Huynh NT, Truong V, Nghia TLH (2023) Vietnamese EFL learners' perspectives on learning English online and employability. In: English language education for graduate employability in Vietnam. Springer Nature Singapore, Singapore, pp 327–351

Massar Associates (2023) Palestinian graduates and the labour market: employers' perspectives. GIZ & Sharek Youth Forum, Ramallah

Melibari GT (2025) Impact of English-medium instruction on graduate employability. Int J Hum 9(3):123–137

Nair PK, Rahim HL, Chelliah S (2023) A bibliometric analysis on the employability of business graduates: a global perspective. J Educ E-Learn Res 10(1):112
Nassar S, Nassar S (2025) Creating an appropriate professional development model for in-service English language teachers in Palestine in light of teachers' needs and global trends
Nghia TLH, Anh NP, Kien LT (2023) English language skills and employability: a theoretical framework. In: English language education for graduate employability in Vietnam. Springer Nature Singapore, Singapore, pp 71–93
Palestinian Ministry of Education (2023) Education and labour market reports. Ramallah
Raymond LM (2025) The lived experiences of English language teachers in conflict-affected settings: a study of the Palestinian context. Doctoral dissertation, Dublin City University
Salim R (2022) Youth unemployment and educational policy in Palestine: challenges and opportunities. J Palest Soc Stud 7(1):30–48. https://www.redalyc.org/journal/5757/575772879005/html/
Ting S-H, Marzuki E, Chuah K-M, Misieng J, Charles J (2017) Employers' views on the importance of English proficiency and communication skill for employability in Malaysia. Indones J Appl Linguist 7(2):315
USAID (2023) Educational mobility and English language proficiency in Palestine. Washington, DC
World Bank (2022) Enhancing job opportunities for youth in MENA: English as a gateway skill. Washington, DC
Zughoul MR (2023) The Arab EFL graduate and the English job market: a study in mismatch. Palest J Lang Soc 12(1):45–63
Zyoud MM, Dweikat KA (2023) Strategies used by in service EFL teachers when teaching English for Palestine Curriculum: case study of Teacher Education Improvement Project. Intern J Advance Appl Res 11(1):251–262. https://doi.org/10.5281/zenodo.10156244

16 The Quality of Digital Services in Palestine in the Context of Governmental Digital Transformation: Literature Review and Exploration of Challenges and Solutions

Eman Naser-Karajah and Lama Subhi Awwad

Abstract

The quality of digital services has become a crucial benchmark for assessing the performance and effectiveness of public institutions in the context of accelerating global digital transformation. This study examines the key challenges and gaps affecting the quality of digital services in Palestine through a descriptive and literature-based analytical review of relevant local and international sources. The analysis identifies major constraints related to digital infrastructure, cybersecurity, institutional coordination, regulatory frameworks, and user experience. The study further emphasizes the influence of the digital divide and insufficient digital literacy on citizens' access to and contentment with governmental digital services. Based on the reviewed literature and policy reports, the paper proposes context-sensitive recommendations aimed at improving digital service quality, strengthening e-governance practices, and enhancing public trust in governmental digital platforms in Palestine.

Keywords

Digital transformation • Quality of digital services • Digital governance • Digital divide • User satisfaction • Transparency

E. Naser-Karajah (✉)
Department of Information Technology, The Arab American University Palestine, Ramallah, Palestine
e-mail: eman.naser@aaup.edu

L. S. Awwad
Ministry of Labor, Ramallah, Palestine
e-mail: lamaawwad75@gmail.com

B. Awwad and R. A. Awwad (eds.), *Technology and Innovation Towards a Future of Business Sustainability*, Contributions to Environmental Sciences & Innovative Business Technology, https://doi.org/10.1007/978-981-95-6469-9_16

16.1 Introduction

In recent years, governments worldwide have accelerated efforts toward digital transformation, particularly in the wake of the COVID-19 pandemic, which forced many public institutions to adopt digital platforms to ensure service continuity (United Nations 2022; Awwad and Dhair 2025). The quality of digital services has consequently emerged as a key indicator of the success of digital transformation initiatives, particularly in resource-limited and politically complex contexts like Palestine (Awwad and Dhair 2025; Ministry of Telecommunications and Information Technology 2021).

Scholarly research indicates that the quality of digital services is influenced by multiple factors, including technological infrastructure, the efficiency of government staff, the regulatory environment, and the extent of citizen engagement in service design (Obaid et al. 2021; Parasuraman et al. 1988; Davis 1989). Other critical dimensions include usability, security, and response time (Parasuraman et al. 1988). In the Arab region, digital transformation initiatives continue to face obstacles such as the digital divide, lack of transparency, and weak institutional governance (ESCWA 2023; Alkhateeb and Abdalla 2021).

In Palestine, despite the launch of platforms like Hakomati for unified E-Government services (Ministry of Telecommunications and Information Technology 2023), there is still a limited scientific evaluation of their quality. This research builds on global and regional literature and established evaluation models, such as SERVQUAL and the E-Government Development Index (EGDI), to examine the status of digital services in Palestine and propose evidence-based recommendations to enhance their effectiveness (Parasuraman et al. 1988; OECD 2020).

16.2 Methodology

This study adopts a descriptive and literature-based analytical approach to examine the state of digital service quality in Palestine through reviewing relevant academic literature and policy documents. Secondary sources include peer-reviewed journal articles, doctoral dissertations, international reports, and national strategies related to digital transformation and digital service delivery (United Nations 2022; Ministry of Telecommunications and Information Technology 2021).

Key sources include the National Digital Transformation Strategy 2021–2025 (Ministry of Telecommunications and Information Technology 2021), the Report on the Status of Digital Services 2023 (Ministry of Telecommunications and Information Technology 2023), and technological surveys from the Palestinian Central Bureau of Statistics (Palestinian Central Bureau of Statistics 2023).

16.3 Theoretical Framework and Literature Review

The theoretical framework underpins the relationship between governmental digital transformation and the quality of digital services. Digital transformation extends beyond the digitization of paper-based processes to encompass the redesign of workflows and services aimed at achieving greater efficiency, transparency, and user-centered service delivery. (Awwad and Dhair 2025; Obaid et al. 2021; Davis 1989; Awwad et al. 2024).

Digital service quality is a multidimensional concept encompassing the ability of E-Government platforms to meet users' expectations efficiently, securely, and with ease of access. Core dimensions include usability, reliability, security, responsiveness, and user satisfaction (Parasuraman et al. 1988; Alkhateeb and Abdalla 2021).

This study employs multiple well-recognized models for evaluating digital service quality, including SERVQUAL for measuring gaps between expectations and actual performance (Parasuraman et al. 1988), the Technology Acceptance Model for understanding users' willingness to adopt technology (Davis 1989), and the EGDI for assessing national digital readiness (United Nations 2022). Models emphasizing User experience were also considered to capture the significance of user-centered design in digital services (Hussein 2019; Abdalla 2023).

Global literature, including reports by the UN and OECD, stresses that successful digital transformation requires strong digital infrastructure, skilled human capital, inclusive access, and supportive legal frameworks (United Nations 2022; OECD 2020). In contrast, Palestinian studies reveal persistent challenges in system integration, service continuity, and user-centered design (Ministry of Telecommunications and Information Technology 2023; Palestinian Central Bureau of Statistics 2023; Abu Mansour 2022; Soudi 2022).

16.4 Challenges Affecting Digital Service Quality in Palestine

Palestine faces multifaceted challenges that hinder the quality and sustainability of its digital services. From a technical perspective, Palestine faces weak digital infrastructure, limited network coverage, and inadequate cybersecurity measures, resulting in unstable and inconsistent digital service quality. (Parasuraman et al. 1988; Ministry of Telecommunications and Information Technology 2023).

Prior studies in the Palestinian public sector emphasize the role of governance principles in enhancing service quality (Awwad et al. 2021). Institutionally, the shortage of qualified staff and insufficient inter-agency coordination impede the delivery of integrated and seamless e-services (Ministry of Telecommunications and Information Technology 2023; Al Najjar et al. 2023).

Legislative gaps further exacerbate these issues, as Palestine lacks comprehensive legal frameworks for digital transformation, data protection, and privacy, which undermines public trust and slows the development of digital services (Ministry of Telecommunications and Information Technology 2021).

Social challenges include a significant digital divide affecting marginalized communities, people with disabilities, and older people, coupled with low digital literacy that limits the effective utilization of available services (ESCWA 2023; Soudi 2022).

Finally, political constraints stemming from the Israeli occupation, including control over telecommunication infrastructure and restricted access to modern technologies, represent major barriers to digital progress (Palestinian Central Bureau of Statistics 2023; Soudi 2022).

16.5 Recommendations and Proposed Solutions

Addressing these challenges requires a coordinated and multifaceted approach. First, upgrading digital infrastructure and expanding network coverage are critical to ensuring stable and secure digital services, alongside strengthening cybersecurity mechanisms (Parasuraman et al. 1988; Ministry of Telecommunications and Information Technology 2023).

Second, strengthening human and institutional capacity is essential. Specialized and continuous training programs for government employees will enhance their technical and managerial competencies and support innovative digital solutions (Davis 1989; Abu Mansour 2022; Awwad 2024).

Third, developing robust legal frameworks for data protection, digital rights, and cybersecurity is necessary to improve governance, transparency, and public trust (Ministry of Telecommunications and Information Technology 2021; Wirtz and Daiser 2018).

Fourth, bridging the digital divide requires inclusive awareness campaigns and targeted programs to empower marginalized groups, women, older people, and people with disabilities (ESCWA 2023; Soudi 2022).

Finally, adopting comprehensive evaluation models such as SERVQUAL and EGDI will enable periodic assessments of service quality and user satisfaction, while leveraging international best practices adapted to Palestine's local context (Parasuraman et al. 1988; OECD 2020; Abdelhamid 2011).

16.6 Discussion and Analysis

The reviewed literature indicates a clear gap between Palestine's digital transformation strategies and their practical implementation. Weak inter-agency coordination, limited resources, and inadequate focus on user experience remain key obstacles (Ministry of Telecommunications and Information Technology 2023; Abu Mansour 2022).

International best practices show that successful digital transformation relies on comprehensive strategies integrating infrastructure upgrades, human capacity development, legal modernization, and inclusive access (United Nations 2022; OECD 2020; Salah and Awwad 2024).

In the Palestinian context, the absence of effective user-centric design and insufficient engagement of vulnerable populations contributes to low satisfaction rates and limited adoption of digital platforms (Soudi 2022; Abdelhamid 2011). Addressing these gaps requires implementing adaptable, user-focused solutions grounded in systematic evaluation and international experience.

16.7 Conclusion

This study concludes that Palestine's digital service quality faces persistent technical, institutional, legislative, and social challenges, compounded by political constraints. Despite partial progress, significant gaps remain in infrastructure, coordination, user experience, and legal frameworks. Developing an integrated national evaluation framework, strengthening cybersecurity, empowering government employees, and adopting inclusive and user-centered design principles are essential steps toward achieving sustainable digital transformation in Palestine. Leveraging international best practices while adapting to local socio-political realities is key to building citizen trust and improving the overall quality of digital services.

References

Abdalla RAM (2023) Key determinants of citizens' behavioural intentions towards using the e-government services' portal in Palestine. Doctoral dissertation

Abdelhamid A (2011) Towards e-municipality in the Palestinian territories: Qalqilia Municipality as a case study. Doctoral dissertation, An-Najah National University

Abu Mansour A (2022) Investigating the readiness of ICT Palestinian organizations for digital transformation. Doctoral dissertation, An-Najah National University

Al Najjar MT, Al Shobaki MJ, Talla E (2023) The reality of digital transformation in non-governmental organizations in Palestine

Alkhateeb MA, Abdalla RA (2021) Antecedents of trust in e-government: Palestinian citizens' perspective. Stud Appl Econ 39(7)

Awwad B (ed) (2024) The AI revolution: driving business innovation and research: volume 1, vol 524. Springer Nature

Awwad BS, Dhair MA (2025) The impact of implementing digital transformation on the quality of government service provision in the Palestinian Ministry of Higher Education and Scientific Research. In: Sustainability in light of governance and artificial intelligence applications, p 71

Awwad B, Al-Shomali S, Salem M (2021) The impact of implementing governance principles on the quality of services in the Palestinian Ministry of Telecommunications and Information Technology. J Added Value Bus Econ 3(1):1–21

Awwad BS, Razia BS, Razia AS (2024) Digital transformation under the governance of Palestinian banks. Discov Sustain 5(1):76

Davis FD (1989) Perceived usefulness, perceived ease of use, and user acceptance of information technology. MIS Q 13(3):319–340

ESCWA (2023) Bridging the digital divide in Arab countries: challenges and recommendations. United Nations, Beirut

Hussein BH (2019) Analysis of citizens' satisfaction (acceptance & needs) for e-government public services: evidence from Palestine. Doctoral dissertation, Arab American University, Palestine

Ministry of Telecommunications and Information Technology (2021) National digital transformation strategy 2021–2025. Ramallah, Palestine

Ministry of Telecommunications and Information Technology (2023) Report on the status of digital services in palestine. Ramallah, Palestine

Obaid T, Eneizan B, Naser SSA, Alsheikh G, Ali AAA, Abualrejal HME, Gazem NA (2021) Factors contributing to an effective e-government adoption in Palestine. In: International conference of reliable information and communication technology. Springer, Cham, pp 663–676

OECD (2020) Digital government index: 2019 results. OECD public governance policy papers, no. 03. OECD Publishing, Paris

Palestinian Central Bureau of Statistics (2023) Technological survey of palestinian institutions. Ramallah, Palestine.

Parasuraman A, Zeithaml VA, Berry LL (1988) SERVQUAL: a multiple-item scale for measuring consumer perceptions of service quality. J Retail 64(1):12–40

Salah AS, Awwad BS (2024) A theoretical review of artificial intelligence and tax compliance. In: Achieving sustainable business through AI, technology education and computer science: volume 3: business sustainability and artificial intelligence applications, pp 53–64

Soudi L (2022) The influence of digital transformation on the quality level of the governmental services: the case of the Palestinian Land and Water Settlement Authority. Doctoral dissertation, Al-Quds Open University

United Nations (2022) E-government survey 2022: the future of digital government. UNDESA, New York

Wirtz BW, Daiser P (2018) Strategic digital transformation management: key factors and impact on performance. J Strat Inf Syst 27(2):139–151

Digital Empowerment and Remote Work in Palestine: Towards an Economic Sustainability Strategy Through Digital Transformation

17

Lama Subhi Awwad and Bahaa Awwad

Abstract

His study examines digital empowerment and remote work in Palestine amid weak infrastructure, limited supportive policies, and movement restrictions. Adopting a descriptive-analytical framework, it analyzes official statistics and field reports to map the digital divide, reviews national legislation on governance, cybersquatting, and remote labor, and benchmarks lessons from comparable fragile contexts. The research connects key digital enabler—cloud computing, automation, and data governance to institutional sustainability goals of cost efficiency, operational flexibility, and social inclusion. The resulting road map prioritizes expanding broadband infrastructure, up-skilling local talent, and enacting supportive national policies. Findings confirm that only an integrated blend of technology, policy, and sound governance, aligned with local market needs, can unlock effective digital empowerment and remote work, bolstering the resilience and growth capacity of Palestinian institutions in an unstable environment.

Keywords

Digital empowerment • Remote work • Digital transformation • Institutional sustainability • Palestine • Digital governance • Cloud computing • Digital legislation • Transformation strategies

L. S. Awwad
Ministry of Labor, Ramallah, Palestine

B. Awwad (✉)
College of Business and Finance, Ahlia University, Manama, Bahrain
e-mail: bawwad@ahlia.edu.bh

B. Awwad and R. A. Awwad (eds.), *Technology and Innovation Towards a Future of Business Sustainability*, Contributions to Environmental Sciences & Innovative Business Technology, https://doi.org/10.1007/978-981-95-6469-9_17

17.1 Introduction

Global technological: shifts have made digital transformation, empowerment, and remote work essential for advancing economic and social sustainability. In the Palestinian context, these tools offer strategic responses to political and economic constraints that limit mobility and access to conventional labor markets.

Research Aim and Scope: This study investigates the current realities of digital empowerment and remote work in Palestine, identifies key obstacles, and suggests viable strategies to support institutional sustainability in a volatile environment.

Methodological Approach: The research employs a descriptive-analytical method, drawing on official data and field reports to evaluate digital infrastructure and map the digital divide. It also reviews national digital legislation and draws on global case studies from fragile contexts to inform localized models.

Institutional Linkages: Key digital tools—such as cloud computing, data governance, and automation—are analyzed for their role in promoting institutional sustainability through cost savings, improved flexibility, and greater inclusion.

Road Map and Key Conclusion: The study proposes a focused action plan to enhance digital infrastructure, up skill the workforce, and align national policy frameworks with digital transformation goals. It concludes that progress requires an integrated, well-governed approach tailored to Palestine's developmental context to strengthen institutional resilience and economic growth.

17.2 Literature Review

17.2.1 Digital Empowerment

Digital empowerment equips individuals and institutions with the infrastructure, devices, and skills needed to leverage ICT for greater productivity and quality of life. Robust broadband, affordable smart devices, and targeted training narrow digital divides, spur innovation, and boost employment and institutional performance (Smith 2020; Johnson and Lee 2019; Al-Masri 2021; World Bank 2024).

17.2.2 Remote Work

Remote work—ranging from fully home-based to hybrid models—relies on digital tools for coordination (Davis 2018; Brown and Smith 2020). It enhances flexibility, work–life balance, and productivity while cutting commuting costs (Green 2021; Miller 2019). Core obstacles include patchy infrastructure, limited user skills, and cybersecurity risks (Khan 2020).

17.2.3 Digital Transformation in Organizations

Digital transformation reshapes all business processes through technologies such as cloud computing, automation, and data analytics to attain a sustainable competitive advantage (Porter and Heppelmann 2015). Effective initiatives pair technical upgrades with human capacity development, innovation culture, and strong digital governance frameworks to ensure compliance, data protection, and risk control (ISACA 2019; Mell and Grance 2011).

17.2.4 Digital Transformation and Economic Sustainability

Digitization underpins sustainability by raising efficiency, reducing resource waste, and enabling scalable, low-cost operations via cloud services (Elkington 1997; Armbrust et al. 2010). Sound data governance strengthens privacy, trust, and continuity (Davenport 2014), while remote work expands labor-market access and helps close socioeconomic gaps (UNDP 2021).

17.2.5 International and Palestinian Experience

Globally—accelerated by COVID-19—countries such as Spain and Canada leveraged cloud platforms to preserve business continuity (OECD 2020; Government of Canada 2021). In Palestine, select education and health institutions have piloted remote-work systems despite movement restrictions and limited bandwidth (PCBS 2022). Scaling success will require cohesive national policy, upgraded infrastructure, and regulatory reforms to overcome persistent technical and legal barriers (Ministry of Telecom 2023).

17.3 Methodology

17.3.1 Research Design

The study adopts a descriptive-analytical design, combining qualitative and limited quantitative elements to examine how digital transformation and remote work affect economic and institutional sustainability in Palestine.

17.3.2 Data Sources and Tools

- Two complementary qualitative tools were used.
- Document Analysis of national statistics (PCBS), ministerial reports, and policy papers from UNDP, World Bank, OECD, and Palestinian universities to map digital trends and challenges.

- Discourse Analysis of government-ministry websites, press releases, and media coverage (2020–2025) to capture narratives on digital transformation and remote work.

17.3.3 Population and Sampling

The target population comprises employees and experts across public, private, and civil-society institutions involved in digital initiatives. Purposive sampling selected documents and discourses that were: (1) relevant to digital empowerment or remote work, (2) credible, (3) published between 2020 and 2025, and (4) representative of multiple sectors.

17.3.4 Procedures

Between March and May 2025, documents were retrieved and coded thematically (infrastructure, governance, barriers, opportunities, sustainability). Discourse materials were processed using NVivo or manual coding to identify language patterns and implicit messages. All sources were logged in a transparent audit trail to ensure replicability.

17.3.5 Scope and Limitations

- Geographical: West Bank, Gaza Strip, East Jerusalem.
- Temporal: 2020–2025.
- Topical: Digital empowerment, remote work, and sustainability.

Key limitations include lack of primary survey data, possible bias in public narratives, and scarce localized studies. Triangulation across data types mitigated these constraints.

17.3.6 Ethical Considerations

All materials were publicly available or officially authorized. Sources were accurately cited, and analyses remained impartial, avoiding the politicization of sensitive issues.

17.4 Analysis and Discussion

The analytical findings reveal a layered and multifaceted reality surrounding digital transformation and remote work in Palestine. Drawing on both documentary and discourse analyses, the study identifies five interrelated dimensions shaping the Palestinian digital landscape: infrastructure, governance, institutional practice, socio-economic inclusion, and systemic barriers.

17.4.1 Digital Infrastructure Gaps and Uneven Access

The most pressing issue is the uneven distribution of digital infrastructure. Urban centers benefit from relatively higher connectivity due to targeted investments, while rural regions—particularly Gaza and Area C—suffer from limited broadband reach and outdated systems. The digital divide is not only geographic but also institutional, as small and medium enterprises (SMEs) and civil society organizations frequently lack access to cloud computing or secure platforms. This infrastructure disparity exacerbates socioeconomic inequalities and restricts innovation and productivity in marginalized areas.

17.4.2 Political Constraints on Connectivity

Israeli control over telecommunication infrastructure imposes systemic barriers to digital development. These include denial or delay in allocating frequencies, obstruction of 4G and 5G network rollouts, and restrictions on importing technical equipment. As a result, Palestinian institutions are unable to integrate seamlessly into regional and global digital ecosystems, placing them at a competitive disadvantage and deterring foreign digital investment.

17.4.3 Fragmentation of National Digital Strategy

While numerous ministries and public bodies have adopted digital initiatives, their efforts lack coherence due to the absence of a national digital governance framework. Content analysis of policy discourse highlights the lack of centralized oversight, which leads to duplicated platforms, inconsistent data systems, and fragmented services. This decentralization weakens the potential benefits of scale and prevents the formation of an integrated e-governance model.

17.4.4 Remote Work: Opportunities and Resistance

The COVID-19 pandemic accelerated the adoption of remote work, especially in education and public administration. However, the shift was largely unplanned and reactive. Most institutions lacked digital preparedness strategies, which led to gaps in quality assurance, service continuity, and equitable access. Technological constraints were compounded by limited digital literacy among staff and the lack of legal frameworks to protect remote workers' rights. Moreover, entrenched workplace cultures—especially in the public sector—remain skeptical of productivity models that rely on virtual supervision. Without institutional mindsets that embrace flexibility and performance-based accountability, remote work risks being relegated to a crisis-only solution. This is consistent with findings on learning readiness and digital collaboration challenges in e-learning environments (Razzaque et al. 2019).

17.4.5 Islands of Innovation: Local Successes

Despite these challenges, the research uncovered several successful models in municipalities, universities, and tech start-ups that have sustainably integrated remote operations and digital services. These cases demonstrate that with visionary leadership, capacity development, and cross-sector partnerships, even under-resourced institutions can achieve meaningful digital advancement. Such models could be scaled and replicated across sectors, provided national policy alignment is secured.

17.4.6 Governance and Digital Modernization

The connection between digital transformation and institutional governance is critical. Digital platforms can dramatically enhance transparency, streamline administrative processes, and increase citizen engagement. However, most Palestinian institutions are still in early stages—focused more on digitizing forms than on building integrated, secure, and smart systems. The absence of national legislation on cybersquatting, data privacy, and digital labor rights weakens institutional trust and deters citizen participation. As highlighted in recent regional work linking governance, sustainability, and AI-driven transformation frameworks (Awwad 2025), effective digital modernization requires more than digitizing services—it demands integrated, intelligent governance systems.

17.4.7 Financial Sustainability and Donor Dependence

While digital platforms have shown potential in reducing operational costs and improving service delivery, many of these systems rely heavily on short-term donor funding. This financial fragility poses risks to the sustainability of digital gains. Long-term success depends on transitioning from project-based funding to strategic national investment in digital infrastructure and maintenance.

17.4.8 Data Governance and Evidence-Based Policy

A key weakness identified is the lack of unified data systems across institutions. This fragmentation inhibits evidence-based policymaking and limits the capacity for data analytics. Encouragingly, sectors such as health and higher education are beginning to adopt smart systems that integrate data for planning, but such practices remain the exception.

17.4.9 Economic Potential of Digital Transformation

The shift toward digital economies creates new employment pathways, particularly through freelancing, online entrepreneurship, and remote service provision. These models are especially promising for youth, women, and marginalized communities. Reports from international bodies confirm Palestine's potential in sectors like software development, e-commerce, and digital content services. Yet unlocking this potential requires coordinated reforms in education, market regulation, and labor law.

17.4.10 Social Impact and Inclusion

Digital empowerment also holds transformative social potential. It allows individuals from remote and under-served areas to participate in national and global labor markets. Women in conservative or isolated settings gain access to work without violating cultural norms. Likewise, youth facing limited local job opportunities can engage in global gig work. These shifts help reduce dependency on physical infrastructure and traditional labor channels.

17.4.11 Structural and Legal Barriers

Despite promise, structural challenges remain formidable. The ongoing Israeli occupation, political division between the West Bank and Gaza, and the absence of comprehensive legislation create a hostile environment for digital transformation. Institutional capacity is further limited by a lack of skilled personnel and underdeveloped training systems. Digital competencies are uneven across sectors, and public decision-makers often lack the tools to lead digital change.

17.5 Conclusion

Palestine's digital transformation remains constrained by foundational issues—political, infrastructural, legal, and cultural. Yet within these constraints lie clear signs of potential. Isolated successes, growing public interest, and sectoral initiatives offer a base for broader transformation. Achieving this, however, requires not just technical fixes, but an integrated vision—one that mobilizes governance, education, investment, and legislation toward a resilient, inclusive digital future.

This study concludes that digital transformation and remote work in Palestine are not just reactive measures to crises like COVID-19, but strategic paths to restructuring the Palestinian economy and enhancing institutional and social sustainability amid deep political and economic challenges.

Palestine's digital infrastructure faces significant gaps caused by Israeli occupation and fragmented domestic policies, hindering the implementation of effective national digital strategies. Although there are successful institutional models and advanced initiatives, these remain isolated and uncoordinated due to the lack of regulatory frameworks, weak digital governance, and skill shortages in both public and private sectors.

Digital transformation goes beyond technical aspects to include cultural, administrative, and legislative shifts, requiring a fundamental change in institutional thinking. For example, adopting remote work permanently demands rethinking monitoring systems, professional relationships, and evaluation methods, which are still underdeveloped.

Despite these challenges, digital transformation shows great potential to expand job opportunities—especially for youth and women—increase institutional efficiency, reduce operational costs, and broaden access to services, making it a national priority rather than a luxury.

17.6 Strategic Recommendations

- Develop a comprehensive national digital transformation strategy supervised by a central authority, linking priorities to the Sustainable Development Goals and involving public, private, and civil society sectors.
- Reform legal and regulatory frameworks related to data protection, cyber-squatting, and remote work to safeguard workers' rights, user privacy, and accountability of digital service providers.
- Invest in expanding digital infrastructure, especially in geographically and economically marginalized areas, and develop robust digital service platforms.
- Strengthen digital human capital through national training programs and integration of digital skills into general, vocational, and higher education curricula.
- Support sustainable digital work models by incentive's successful institutions financially, regularity, and technically, while documenting and sharing best practices.

- Ensure digital equity and inclusion by prioritizing women, people with disabilities, and residents of under-served areas with accessible and affordable digital solutions.

17.7 Closing Remarks

Digital transformation and remote work offer Palestine a real opportunity to shift toward more efficient, equitable, and resilient development model. Achieving this requires political will, institutional coordination, and long-term investment in people and infrastructure. Without these, digitization will have limited impact and fail to drive the urgent sustainable development Palestine needs.

References

Al-Masri S (2021) Digital empowerment in developing countries. J ICT Dev 12(3):45–62
Armbrust M, Fox A, Griffith R, Joseph AD, Katz R, Konwinski A et al (2010) A view of cloud computing. Commun ACM 53(4):50–58
Awwad BS (ed) (2025) Sustainability in light of governance and artificial intelligence applications. Emerald Group Publishing
Brown J, Smith K (2020) Remote work models and outcomes. Int J Bus Stud 15(2):120–135
Davenport TH (2014) Big data at work: dispelling the myths, uncovering the opportunities. Harvard Business Review Press
Davis R (2018) The rise of remote work. Work Trends Q 9(1):30–44
Elkington J (1997) Cannibals with forks: the triple bottom line of 21st century business. Capstone Publishing
Government of Canada (2021) Digital government strategy report
Green L (2021) Benefits of telecommuting. J Hum Resour 22(4):210–223
ISACA (2019) Digital governance frameworks
Johnson M, Lee S (2019) Closing the digital divide. Soc Incl J 8(1):10–25
Khan A (2020) Cybersecurity challenges in remote work. J Cybersecur 5(3):60–75
Mell P, Grance T (2011) The NIST definition of cloud computing. National Institute of Standards and Technology
Miller T (2019) Employee satisfaction in remote work settings. J Organ Behav 40(2):98–112
OECD (2020) Remote work and COVID-19: policy responses and trends
Palestinian Central Bureau of Statistics (PCBS) (2022) ICT usage in Palestine
Palestinian Ministry of Telecom (2023) National digital strategy report
Porter ME, Heppelmann JE (2015) How smart, connected products are transforming competition. Harv Bus Rev 93(10):96–114
Razzaque A, Al Hashimi M, Musleh A, Ahlam H (2019) Learning readiness when sharing knowledge while e-learning. Opción Rev Cienc Hum Soc (19):2930
Smith J (2020) Digital literacy and empowerment. Technol Soc 14(1):5–20
UNDP (2021) Digital inclusion and social equity
Various Local News Reports (2020–2025) Addressing digital empowerment and remote work in Palestine
World Bank (2024) Enhancing the digital economy in the occupied Palestinian territories. Washington, USA

MIX
Papier aus verantwortungsvollen Quellen
Paper from responsible sources
FSC® C105338

If you have any concerns about our products,
you can contact us on
ProductSafety@springernature.com

In case Publisher is established outside the EU,
the EU authorized representative is:
Springer Nature Customer Service Center GmbH
Europaplatz 3, 69115 Heidelberg, Germany

Printed by Libri Plureos GmbH
in Hamburg, Germany